GUIDE TO INFORMATION SOURCES IN THE CONSTRUCTION INDUSTRY

Compiled by
JULES B. GODEL

Construction Publishing Company Inc.

GUIDE TO INFORMATION SOURCES IN THE CONSTRUCTION INDUSTRY

Library of Congress Catalog Card Number:
73-78673.
ISBN: 0-913634-12-3.

Construction Publishing Company, Inc.
Two Park Avenue, New York, New York 10016.

PREFACE

GUIDE TO INFORMATION SOURCES IN THE CONSTRUCTION INDUSTRY is the outgrowth of a two-year effort by a group of compilers whose purpose it was to contact every major organization that publishes construction information whether it be pamphlet or encyclopedia, and to assemble summaries of the information and how and where it could be obtained. More than 13,000 books, standards, reports, etc., were identified.

It soon became obvious to us that there was a need to publish a streamlined version of our larger data base that would give a rundown of the myriad organizations that comprise the construction industry, and to include their major publications and reference sources. The advantages of such a book are obvious; it can be updated frequently to keep pace with the fast-changing industry. The reader can easily contact the organizations listed to gain specific information—and ultimately, practically all of the data of the larger data base can be made available. Finally, this book is less voluminous and designed to give you the most information for the least cost.

For information in the particular areas of interest captioned below, the appropriate source is listed.

1. BOOKS *Scientific & Technical Books in Print*
   R. R. Bowker Company
   1180 Avenue of the Americas
   New York, New York 10036
   (Published Annually)

2. GOVERNMENT REPORTS
   2.1 NATIONAL TECHNICAL INFORMATION SERVICE (NTIS)
   Springfield, Virginia 22151

a. *Government Reports Announcements:* Published Semimonthly @ $52.50/yr
(Abstracts of Govt. research reports)

b. *Weekly Government Abstracts;* Newsletter

Civil & Structural Engineering;	$25/yr
Building Technology	$17.50/yr
Transportation	$17.50/yr
Urban Technology	$35/yr

2.2 GOVERNMENT PRINTING OFFICE
Washington, D.C. 20402
Monthly Catalog of U.S. Government Publications

3. STANDARDS & CODES

3.1 AMERICAN NATIONAL STANDARDS INSTITUTE (ANSI)
1430 Broadway, New York, New York 10018
Standards Catalog: Published Annually, no charge
More than 550 construction standards prepared by about one dozen professional and trade groups are listed for sale.

3.2 MODEL BUILDING CODES
There are five organizations that publish almost 40 model codes on housing, property, one- and two-family dwellings, building, fire, electrical, mechanical and plumbing.

These are:
Building Officials Conference of America (BOCA)
International Conference of Building Officials (ICBO)
Southern Building Code Congress (SBCC)
American Insurance Association (AIA)
National Fire Protection Association (NFPA)

See listings in Section V for addresses.

The author wishes to thank Mr. Harvey A. Harkaway, publisher of *Business Publication Rates and Data,* a publication of

Standard Rate and Data Services, Inc. for permission to use descriptive material of many periodicals. Thanks also to George Watson for his excellent job of editing this book and for extracting some of the salient data from the larger store of construction industry information.

Jules B. Godel

CONTENTS

INTRODUCTION

This book is intended to be useful to the breadth of the construction industry, from the entrepreneur or government official who conceives a project to the manufacturer of ornamental hardware who finishes it; and to the architects, landscapers, consultants, electrical engineers, quarriers, building managers, labor unions, lumber merchants—the spectrum of those who advise, regulate, build, service, and supply.

The book is composed of six sections: Owners; Design Professionals (the technical experts); Materials and Manufacturers (those who supply the basic ingredients); Contractors (those who construct); Standards Groups, Government Planners, etc. (those who plan, control, and regulate); and Other Interests (labor, real estate, etc.). In each section, the reader will find lists of the professional associations in the field, of the major publications (the official newsletters, the technical journals, the news magazines), and of the directories (of members, suppliers, agencies, libraries, and others).

These information sources can be used in a variety of ways. A general contractor entering a specialized field such as hospital construction for the first time can find the experts he needs for help. A firm contemplating building a manufacturing plant in a new area can find the local agencies needed for advice on building codes. A manufacturer introducing a new building material can find the potential customers he seeks among the directories of architects and engineers.

The uses are virtually limitless.

I. OWNERS

In this section, the reader will find lists of owner-oriented organizations, periodicals, and directories. The organizations plan and build—or are otherwise associated with—such far-ranging construction projects as apartment buildings, prisons, hospitals, golf courses, and recreational facilities. The periodical and directory lists contain publications of official associations and those of independent publishers.

APARTMENTS AND HOUSING

Associations and Professional Groups

APARTMENT OWNERS AND MANAGERS ASSOCIATION OF AMERICA, INC., 65 Cherry Plaza, Watertown, Connecticut 06795. Tel.: (203) 274-2589. Composed principally of builders who manage and operate the apartment buildings they construct. Publications:

Newsletter, monthly
Who Is Who in Multi-Family Housing, annual

NATIONAL APARTMENT ASSOCIATION, 1825 K. St., N.W., Suite 1120, Washington, D.C. 20006. Tel.: (202) 785-5111. State and local chapters of managers, owners, and builders of apartments. Publication:

Apartment Profits magazine

NATIONAL ASSOCIATION OF HOUSING COOPERATIVES, Suite 1100, 1828 L. St., N.W., Washington, D.C. 20036. Tel.: (202) 872-0550. Comprises about 90 cooperatives. Publications:

Journal of Cooperative Housing, annual
What's Happening, monthly

Periodicals

APARTMENT NEWS, published monthly by Apartment News Publications, Inc., 4120 Atlantic Ave., Long Beach, California 90807. For builders of apartments, owners, developers, professional managers and professional management firms. Carries news of construction, events affecting the industry, and new products used in the industry. No cost to qualified readers; $4.50 per year for others.

APARTMENT PROFITS, official publication of the National Apartment Association. Published bimonthly by Media America, 2627 Kipling, Houston, Texas 77006. For the apartment owner who builds and operates multi-family housing. Articles concern project feasibility, site selection, building design, methods of construction and materials, solving zoning problems, financing, and apartment management and marketing. New products used in apartment construction, replacement, and remodeling are also featured. No cost to qualified readers; $3.50 per year for others.

BUILDINGS, published monthly by Stamats Publishing Co., 427 Sixth Ave., S.E., Cedar Rapids, Iowa 52406. For those in the commercial and apartment building industry who assume ownership functions and responsibilities. Written from a practical viewpoint toward construction, modernization, operation and maintenance costs are emphasized. No cost to qualified readers; $15 per year for others.

HOUSE & HOME, published monthly by McGraw-Hill Publications, 1221 Avenue of the Americas, New York, N.Y. 10020. For housing professionals whose combined skills in planning and architecture, building methods and technology, land buying and development, finance, management, and marketing are utilized in housing and other light construction markets. $6 per year.

COMMERCIAL AND INDUSTRIAL BUILDINGS

Associations and Professional Groups

AMERICAN BANKERS ASSOCIATION, 1120 Connecticut Ave., N.W., Washington, D.C. 20036. Members comprise more than 95 percent of U.S. banks and trust companies. Tel.: (202) 467-4000. Publication: *Banking,* monthly

AMERICAN HOTEL AND MOTEL ASSOCIATION, 888 Seventh Ave., New York, N.Y. 10019. Federation of state and regional hotel associations representing almost 8,000 hotels and motels. Tel.: (212) 265-4506. Publications:

Buyer's Guide, annual
Directory of Hotel and Motel Systems, annual
Construction and Modernization Equipment, monthly
Product News, semiannual

AMERICAN WATER WORKS ASSOCIATION, INC., 2 Park Ave., New York, N.Y. 10016. Tel.: (212) 686-2040. Professional group of engineers, water works managers, chemists, etc., who develop standards in water supply and water works design and construction. Publication:

Journal, monthly

ATOMIC INDUSTRIAL FORUM, INC., 475 Park Ave. S., New York, N.Y. 10016. Tel.: (212) 725-8300. Industrial firms and labor and educational groups who are engaged in the development and applications of nuclear energy. Publications:

Newsletter
Nuclear Industry, monthly

BUILDING OWNERS AND MANAGERS ASSOCIATION INTERNATIONAL, 224 S. Michigan Ave., Chicago, Illinois 60604. Tel.: (312) 922-0210. For owners and managers of office and apartment buildings. Publication:

Skyscraper Management, monthly

EDISON ELECTRIC INSTITUTE, 90 Park Ave., New York, N.Y. 10016. Tel.: (212) 573-8700. Principal association of investor-owned electric utility companies.

INDEPENDENT BANKERS ASSOCIATION OF AMERICA, P.O. Box 267, Sauk Center, Minnesota 56378. Tel.: (612) 352-2279. Organization of small- to medium-sized banks. Publication:

The Independent Banker, monthly

INTERNATIONAL COUNCIL OF SHOPPING CENTERS, 445 Park Ave., New York, N.Y. 10022. Tel.: (212) 421-8181. Owners, developers and contractors who build or serve the shop-

ping center industry. Publication:
Newsletter, monthly

MOTEL ASSOCIATION OF AMERICA, 122½ East High St., Jefferson City, Missouri 65101. Tel.: (314) 635-3537. Publication:
MAA Newsletter and Washington Report, monthly

NATIONAL ASSOCIATION OF INDUSTRIAL PARKS, Suite 912, 1601 N. Kent St., Arlington, Virginia 22209. Tel.: (703) 525-5638. Builders, developers, and owners of industrial parks. Publications:
Developers Diary, quarterly
Newsletter, monthly

NATIONAL ASSOCIATION OF INDUSTRIAL PLANTS, INC., 300 W. 43rd St., New York, N.Y. 10036. Tel.: (212) 246-2782.

NATIONAL ASSOCIATION OF THEATER OWNERS, INC., 1501 Broadway, New York, N.Y. 10036. Tel.: (212) 594-3325. Owners and operators of motion picture theaters. Publications:
NATO Merchandiser, monthly
NATO Theater Management Digest, monthly

NATIONAL PETROLEUM REFINERS ASSOCIATION, 1725 DeSales St., N.W., Washington, D.C. 20036. Tel.: (202) 638-3722.

NATIONAL RESTAURANT ASSOCIATION, 1530 North Lake Shore Dr., Chicago, Illinois 60610. Tel.: (312) 787-2525. Operators of restaurants, cafeterias, and institutional feeding organizations. Publications:
NRA News, monthly
Washington Report, weekly
Membership Directory

Periodicals

BUILDING OPERATING MANAGEMENT, published monthly by Trade Press Publishing Co., 407 E. Michigan St., Milwaukee, Wisconsin 53201. For operating management including administrative and middle management in commercial, industrial, institutional, and educational buildings, responsible for the operation and replacement of the physical property. No cost to qualified readers.

BUILDINGS, published monthly by Stamats Publishing Co., 427 Sixth Ave., S.E., Cedar Rapids, Iowa 52406. For those in the commercial and apartment building industry who assume ownership functions and responsibilities. Written from a practical viewpoint toward construction, modernization, operation, and maintenance. No cost to qualified readers; $15 per year for others.

INDUSTRIAL CONSTRUCTION, published 10 times a year by G. & M. Publishing Co., 3868 Carnegie Ave., Cleveland, Ohio 44113. For those individuals at architect, architectural/engineering, consulting engineer, and constructor companies and the industrial building departments of companies, who design, specify, and buy products used in industrial construction projects. No cost to qualified readers; $6.50 per year for others.

MODERN STORES AND OFFICES, published alternate months by B. J. Martin Co., Inc. 20 N. Wacker Dr., Chicago, Illinois 60608. Covers general, exterior, and display lighting; comfort heating; electric signs, store fronts, and entrances; air conditioning; electric wiring; displays; snow melting; water heating. $6 per year.

SKYSCRAPER MANAGEMENT, published monthly by Building Owners and Managers Association International, Suite 519, 224 S. Michigan Avenue, Chicago, Illinois 60604. News articles on construction and modernization, economies in operating and maintaining buildings, including improved methods and services. Reviews of new products, materials, equipment and literature, developments in automation. *Directory*—June issue. $5 per year.

Directories

DIRECTORY OF SHOPPING CENTERS IN THE U.S. AND CANADA, published by the National Research Bureau, Inc., 424 N. Third St., Burlington, Iowa 52601. Lists more than 15,000 U.S. and Canadian shopping centers, along with their management, operating, and construction costs and other physical and business data. $70.

PLANT ENGINEERING DIRECTORY & SPECIFICATIONS CATALOG, published annually by Technical Publishing Co., 1301 S. Grove Ave., Barrington, Illinois 60010. Gives manufacturer and supplier catalogs and product specifications for more than 15,000 references. Over 8,600 names, addresses, and phone numbers of local sales offices for plant engineering and maintenance products. $25.

HOSPITALS

Associations and Professional Groups

AMERICAN ACADEMY OF MEDICAL ADMINISTRATORS, 6 Beacon St., Boston, Massachusetts 02108. Publication:
AAMA Executive, quarterly

AMERICAN ANIMAL HOSPITAL ASSOCIATION, Box 1304, 405 South St., Elkhart, Indiana 46514. Tel.: (219) 293-2533. Organization for veterinarians who own animal hospitals. Publications:
Animal Hospital, bimonthly
Directory of Members

AMERICAN ACADEMY OF HEALTH ADMINISTRATION, Box 2243, Rockville, Maryland 20852. Tel.: (304) 725-7683. Health administration and public health. Publications:
Newsletter
Directory

AMERICAN ASSOCIATION OF MEDICAL CLINICS, 719 Prince St., Box 949, Alexandria, Virginia 22313. Tel.: (703) 549-5767. Publication:
Directory of Member Clinics

AMERICAN COLLEGE OF HOSPITAL ADMINISTRATORS, 840 N. Lake Shore Dr., Chicago, Illinois 60611. Tel.: (312) 944-0544. Publications:
Directory, biennial
Hospital Administration, quarterly

AMERICAN HOSPITAL ASSOCIATION, 840 N. Lake Shore Dr., Chicago, Illinois 60611. Tel.: (312) 645-9400. Hospital administrators, department heads, and others responsible for pro-

viding patient care, preparing and implementing administrative and technical methods and procedures, and supervising construction planning. Periodical:
Hospitals, twice monthly

AMERICAN NURSING HOME ASSOCIATION, 1200 15th St., N.W., Washington, D.C. 20005. For owners, operators, and administrators of nursing homes, convalescent homes, and homes for the aged offering skilled nursing care. Publication:
Modern Nursing Home, monthly

AMERICAN SOCIETY FOR HOSPITAL ENGINEERS (OF AHA), 840 N. Lake Shore Dr., Chicago, Illinois 60611. Tel.: (312) 645-9592. Publications:
Directory, annual
Hospital Engineering Newsletter, bimonthly

FEDERATION OF AMERICAN HOSPITALS, INC., Tower Building, Little Rock, Arkansas 72201. Tel.: (501) 376-6818. Organization of privately owned hospitals.

NATIONAL ASSOCIATION OF STATE MENTAL HEALTH PROGRAM DIRECTORS, Bellevue Hotel, 15 E St., S.E., Washington, D.C. 20001. Tel.: (202) 638-0900. Officials in charge of state mental health programs.

Periodicals

HOSPITALS, published twice a month by the American Hospital Association, 840 N. Lake Shore Dr., Chicago, Illinois 60611. For member and nonmember hospitals, administrators, department heads, and other hospital and related persons responsible for providing patient care, preparing and implementing administrative and technical methods and procedures, and supervising construction planning. No cost to qualified readers; $10 per year for others.

HOSPITAL WORLD, published monthly by McKnight Medical Communications, Inc., 550 Frontage Rd., Northfield, Illinois 60093. For those making decisions in hospitals, larger nursing homes, and medical clinics. News and analysis of major developments in patient care, administration, construction, government regulation, new products, purchasing, research, and related areas affecting key hospital departments. No cost to qualified readers.

MODERN HOSPITAL, published monthly by McGraw-Hill Publications, 230 W. Monroe St., Chicago, Illinois 60606. For hospital administrators and department heads. News, comment, and feature articles about patient care, special departments such as operating rooms, central service, pharmacy, and emergency, hospital business operation, law, administrative problems, labor and personnel problems, finance, planning and construction, feeding, housekeeping, and maintenance. No cost to qualified readers; $15 per year for others. *Special issues:* Design and Modernization, Security and Safety.

MODERN NURSING HOME, official publication of the American Nursing Home Association. Published monthly by McGraw-Hill Publications, 230 W. Monroe St., Chicago, Illinois 60606. For doctors, nurses, and lay persons who either own, operate, or administer nursing homes, convalescent homes, and homes for the aged offering skilled nursing care. Discusses design, construction, furnishings, and equipment; management and finances; nursing services; medicine; maintenance and housekeeping; menus and food service; legal responsibilities. No cost to qualified readers; $12 per year for others.

NURSING HOMES, published monthly by Cogswell House, Inc., 222 Wisconsin Ave., Lake Forest, Illinois 60045. Covers management, long-term care, restorative therapy, activities, food service, housekeeping and maintenance, architecture and construction, design and furnishing. No cost to qualified readers; $10 for others.

Directories

GUIDE ISSUE—HOSPITALS, published by the American Hospital Association, 840 N. Lake Shore Dr., Chicago, Illinois 60611. Gives health care data for planning purposes. More than 7,000 U.S. hospitals are listed, along with important management data, including administrators' names. $12.50.

HILL-BURTON PROJECT REGISTER, published by U.S. Department of Health, Education, and Welfare; available from National Technical Information Service, Department of Commerce, Springfield, Virginia 22151. Directory of more than 10,000 Hill-Burton hospital projects. Construction cost data is included along with a physical description of each facility. Order No. PB 198567; $3.

LIBRARIES AND PUBLIC BUILDINGS

Associations and Professional Groups

AMERICAN ASSOCIATION OF MUSEUMS, 2233 Wisconsin Ave., N.W., Washington, D.C. 20007. Tel.: (202) 338-5300. Publications:
Museum Directory
Museum News, monthly

AMERICAN LIBRARY ASSOCIATION, 50 East Huron St., Chicago, Illinois 60611. Tel.: (312) 944-6780. Publishes books and pamphlets on library buildings and equipment.

AMERICAN SOCIETY FOR PUBLIC ADMINISTRATION, 1225 Connecticut Ave., N.W., Washington, D.C. 20036. Tel.: (202) 785-3255. Professional society of government administrators. Publications:
Public Administration News, monthly
Public Administration Review, bimonthly

AMERICAN SOCIETY OF PLANNING OFFICIALS, 1313 E. 60th St., Chicago, Illinois 60637. Tel.: (312) 324-3400. Publications:
ASPO Magazine, monthly
Land Use Control, annual

CATHOLIC LIBRARY ASSOCIATION, 461 W. Lancaster Ave., Haverford, Pennsylvania 19041. Tel.: (215) 649-5250.

INTERNATIONAL ASSOCIATION OF AUDITORIUM MANAGERS, 111 E. Wacker Dr., Chicago, Illinois 60601. Tel.: (312) 664-6610. Operates consulting board to help communities plan new facilities and will consult with architects on specific designs. Publications:

Auditorium News, monthly
Convention Program and Directory, annual

Directory

AUDITORIUM/ARENA/STADIUM GUIDE AND INTERNATIONAL DIRECTORY, published annually by Billboard Publications, Inc., 1719 West End Ave., Nashville, Tennessee 37203. Directory of U.S. and Canadian arenas, auditoriums, and stadiums, along with physical data. No cost to qualified readers; $20 per copy for others.

SCHOOLS

Associations and Professional Groups

AMERICAN ASSOCIATION OF COMMUNITY AND JUNIOR COLLEGES, One Dupont Circle, N.W., Washington, D.C. 20036. Tel.: (202) 293-7050. Statistics on enrollments, finances, etc. Publication:
Community-Junior College Journal, monthly

AMERICAN ASSOCIATION OF DENTAL SCHOOLS, 1625 Massachusetts Ave., N.W., Washington, D.C. 20036. Tel.: (202) 667-9433. Publication:
Journal of Dental Education, monthly

AMERICAN ASSOCIATION OF SCHOOL ADMINISTRATORS, 1801 N. Moore St., Arlington, Virginia 22209. Tel.: (703) 528-0700. Professional association of school administrators. Publications:
Hot Line, 13 times per year
The School Administrator, 13 times per year

AMERICAN ASSOCIATION OF STATE COLLEGES AND UNIVERSITIES, One Dupont Circle, N.W., Washington, D.C. 20036. Tel.: (202) 293-7070. Publication:
Memo to the President, semimonthly

ASSOCIATION OF AMERICAN COLLEGES, 1818 R St., N.W. Washington, D.C. 20009. Tel.: (202) 265-3137. Publication:
Liberal Education, quarterly

ASSOCIATION OF AMERICAN MEDICAL COLLEGES, One Dupont Circle, N.W., Washington, D.C. 20036. Tel.: (202) 466-5100. Publication:
Journal of Medical Education, monthly

ASSOCIATION OF AMERICAN UNIVERSITIES, One Dupont Circle, N.W., Washington, D.C. 20036. Tel.: (202) 293-6177. Presidents of American Universities.

ASSOCIATION OF PHYSICAL PLANT ADMINISTRATION OF UNIVERSITIES & COLLEGES, One Dupont Circle, N.W., Washington, D.C. 20036. Tel.: (202) 785-3062. Publication:
Newsletter, monthly

NATIONAL ASSOCIATION OF COLLEGE AND UNIVERSITY ADMINISTRATORS, 1201 16th St., N.W., Washington, D.C. 20036. Tel.: (202) 833-4261.

NATIONAL COUNCIL OF INDEPENDENT COLLEGES AND UNIVERSITIES, One Dupont Circle, N.W., Washington, D.C. 20036. Tel.: (202) 293-1245. Publication:
Annual Directory

NATIONAL COUNCIL OF INDEPENDENT JUNIOR COLLEGES, One Dupont Circle, N.W., Washington, D.C. 20036. Tel.: (202) 293-7050.

NATIONAL SCHOOL BOARDS ASSOCIATION, State National Bank Plaza, Evanston, Illinois 60201. Tel.: (312) 869-7730. Publications:
American School Board Journal, monthly
State Association Directory

MISCELLANEOUS OWNERS

Associations and Professional Groups

AMERICAN ASSOCIATION OF HOMES FOR THE AGING, 374 National Press Bldg., Washington, D.C. 20004. Tel.: (202) 347-2000. Publications:
AAHA News Scene, quarterly
Directory of Non-Profit Homes for the Aged

AMERICAN CEMETERY ASSOCIATION, 250 E. Broad St., Columbus, Ohio 43215. Tel.: (614) 221-6829. Managers of public, private, and religious cemeteries. Publications:
Cemetery Directory, annual
The Cemeterian, monthly

AMERICAN COLLEGE OF NURSING HOME ADMINISTRATORS, 1641 Colesville Rd., Suite 409, Silver Spring, Maryland 20910. Tel.: (301) 589-9070. Publication:
ACNHA Newsletter, bimonthly

AMERICAN CORRECTIONAL ASSOCIATION, 4321 Hartwick Rd., Suite 208, College Park, Maryland 20740. Tel.: (301) 864-1070. Publications:
American Journal of Correction, bimonthly
Directory of State and Federal Correctional Institutions and Agencies

AMERICAN LAND DEVELOPMENT ASSOCIATION, 1000 16th St., N.W., Washington, D.C. 20036. Tel.: (202) 659-4582. Recreational housing and land development: builders and manufacturers. Publications:
American Land, quarterly
Washington Developments, monthly
Directory

ASSOCIATION OF STATE CORRECTIONAL ADMINISTRATORS, Illinois Department of Corrections, 160 N. La Salle St., Chicago, Illinois 60601. Tel.: (217) 525-7263. Administrators of state and federal institutions.

CHAMBER OF COMMERCE OF THE U.S., 1615 H. St., N.W., Washington, D.C. 20006. Tel.: (202) 659-6000. Publications:
Chamber of Commerce Newsletter, monthly
Nation's Business, monthly

GOLF COURSE SUPERINTENDENTS ASSOCIATION OF AMERICA, 3158 Des Plaines Ave., Des Plaines, Illinois 60018. Tel.: (312) 824-6147. Publications:
The Golf Superintendent, 10 times per year
Membership Directory

HARNESS TRACKS OF AMERICA, INC., 333 N. Michigan Ave., Chicago, Illinois 60601. Tel.: (312) 332-1846. Publication:
Directory

INTERNATIONAL ASSOCIATION OF CHIEFS OF POLICE, 11 Firstfield Rd., Gaithersburg, Maryland 20760. Tel.: (301) 948-0922. Publications:
The Police Chief, monthly
The Police Yearbook

INTERNATIONAL ASSOCIATION OF FAIRS AND EXPOSITIONS, 500 Ashland Ave., Chicago Heights, Illinois 64011. Tel.: (312) 756-2029. Publication:
Fairs and Expositions, monthly

INTERNATIONAL ASSOCIATION OF REHABILITATION FACILITIES, 5530 Wisconsin Ave., Suite 955, Washington, D.C. 20015. Tel.: (202) 654-5882. For those concerned with providing facilities for the handicapped. Guidance for architects. Publication:
Focus on Facilities, bimonthly

NATIONAL ASSOCIATION OF STATE DEVELOPMENT AGENCIES, 2000 K. St., N.W., Suite 751, Washington, D.C. 20006. Tel.: (202) 337-1881.

NATIONAL RECREATION AND PARK ASSOCIATION, 1601 N. Kent St., Arlington, Virginia 22209. Tel.: (703) 525-0606. Publications:
Journal of Leisure Research, quarterly
Parks and Recreation, monthly

Periodical

ALASKA INDUSTRY, published monthly by Alaska Industrial Publications, Inc., 409 W. Northern Lights Blvd., P.O. Box 399, Anchorage, Alaska 99510. For management personnel in Alaska business and industry, and executives in other areas who are particularly interested in developments in Alaska. Covers the fields of oil and gas, transportation, construction, mining, timber, fisheries, and general business in Alaska and the closely adjoining area of northwestern Canada. Special issue in February: Businessman's Guide to Alaska. No cost to qualified readers; $20 per year for others.

II. DESIGN PROFESSIONALS

This section guides the reader to the experts in such technical specialties as reinforced concrete, acoustics, paint, earthquakes—to name only a few. The reader will also find experts in human-oriented fields—food service facilities, store planning, and social concerns such as the aging. In the *Publications* part of this section is an extensive listing of the journals the design experts read. And in the *Directories* part are featured publications containing membership lists of the design professionals' organizations—from individual state professional engineers' societies to such large national organizations as the Institute of Electrical and Electronics Engineers. Also featured in *Directories* are catalogs of the nation's engineering and architectural libraries.

Associations and Professional Groups

AMERICAN ACADEMY OF CONSULTANTS, 225 E. 60th St., New York, N.Y. 10022. Tel.: (212) 759-1089. Publication:
Academy Forum, quarterly

AMERICAN ACADEMY OF ENVIRONMENTAL ENGINEERS, Box 1278, Rockville, Maryland 20850. Tel.: (301) 762-7797. Members are required to be licensed and to have more than eight years' experience. Publications:
The Diplomate, quarterly
AAEE Roster, annual

AMERICAN ASSOCIATION OF COST ENGINEERS, 308 Monongahela Bldg., Morgantown, West Virginia 26505. Tel.: (304) 296-8444. Publication:
AACE Bulletin, bimonthly

AMERICAN ASSOCIATION OF HOSPITAL CONSULTANTS, 1700 K St., N.W., Washington, D.C. 20006. Tel.: (202) 785-3434. Organization of experts in hospital planning and administration. Periodical:

Directory, annual

AMERICAN CHEMICAL SOCIETY, 1155 16th St., N.W., Washington, D.C. 20036. Tel.: (202) 737-3337. Publications:

Chemical and Engineering News, weekly
Chemical Technology, monthly
Environmental Science and Technology, monthly

AMERICAN CERAMIC SOCIETY, 65 Ceramic Dr., Columbus, Ohio 43234. Tel.: (614) 268-8645. Professional society for those interested in glass, refractories, whiteware, and structural clay. Publications:

Journal of the ACS, monthly
ACS Bulletin, monthly

AMERICAN CONFERENCE OF GOVERNMENTAL INDUSTRIAL HYGIENISTS, P.O. Box 1937, Cincinnati, Ohio 45201. Association of practicing industrial hygiene personnel in federal, state, and local agencies. Assembles guides and practical information for the evaluation and control of the industrial environment.

AMERICAN INSTITUTE OF ARCHITECTS, 1735 New York Ave., N.W., Washington, D.C. 20006. Tel.: (202) 265-3113. Professional society of more than 22,000 architects. Publications:

AIA Memo, biweekly
AIA Journal, monthly

AMERICAN INSTITUTE OF BUILDING DESIGN, 839 Mitten Rd., Suite 128, Burlingame, California 94010. Tel.: (415) 697-7680. Fosters high standards of building design through consultation with members, annual awards for excellence in design, and through contact with law-making bodies. Publications:

Newsletter, semimonthly
American Journal of Building Design, monthly
Western Building Design, monthly

AMERICAN INSTITUTE OF CHEMICAL ENGINEERS, 345 E. 47th St., New York, N.Y. 10017. Tel.: (212) 752-6800. Seven divisions: Food and Bio-Engineering, Forest Products, Heat Transfer and Energy Conversion, Materials Engineering and Sciences, Nuclear Engineering, Petroleum and Petrochemical, and Environmental. Publications:
Chemical Engineering Progress, monthly
AIChE Journal, bimonthly
International Chemical Engineering

AMERICAN INSTITUTE OF CONSULTING ENGINEERS, 345 E. 47th St., New York, N.Y. 10017. Tel.: (212) 752-6800. Professional society of consultants in the fields of mechanical engineering, civil engineering, electrical engineering, urban planning, etc. Publication:
Consulting Engineering, monthly

AMERICAN INSTITUTE OF PLANNERS, 1776 Massachusetts Ave., N.W., Washington, D.C. 20036. Tel.: (202) 872-0611. Professional organization for those engaged in planning on city, county, state, regional, or national levels. Publications:
AIP Newsletter, monthly
Journal of the AIP, bimonthly
Planners' Notebook, bimonthly

AMERICAN INSTITUTE OF PLANT ENGINEERS, 1021 Delta Ave., Cincinnati, Ohio 45208. Tel.: (513) 321-9412. For plant engineers who design and operate industrial facilities. Publications:
AIPE Newsletter, monthly
Plant Engineer's Guide to the Literature on Air and Water Pollution

AMERICAN NUCLEAR SOCIETY, 244 E. Ogden Ave., Hinsdale, Illinois 60521. Tel.: (312) 325-1911. Publications:
Nuclear News, monthly
Nuclear Technology, monthly
Nuclear Science & Engineering, monthly
Remote Systems Technology Proceedings, annual

AMERICAN SOCIETY FOR ENGINEERING EDUCATION, One Dupont Circle, N.W., Suite 400, Washington, D.C. 20036. Tel.: (202) 293-7080. Professional Society of Engineering educators. Publications:

Engineering Education, 8 times per year
Directory of Engineering College Research and Graduate Study, annual

AMERICAN SOCIETY FOR METALS, Metals Park, Ohio 44073. Tel.: (216) 338-5151. Professional organization for metallurgists and engineers. Publications:

ASM News Monthly
Metal Progress, monthly
Metals Engineering, quarterly
Metals Handbook

AMERICAN SOCIETY FOR NONDESTRUCTIVE TESTING, 914 Chicago Ave., Evanston, Illinois 60202. Metallurgists, welding engineers, and equipment manufacturers. Fosters information on radiography, ultrasonic, dye-penetrant, and magnetic testing techniques. Publication:

Materials Evaluation, monthly

AMERICAN SOCIETY FOR TESTING AND MATERIALS, 1916 Race St., Philadelphia, Pennsylvania 19103. Establishes voluntary standards for numerous materials, products, and test methods, and sponsors research projects on a wide spectrum of materials-related subjects. Publications:

ASTM Journal of Testing and Evaluation, bimonthly
ASTM Standardization News, monthly

AMERICAN SOCIETY OF CIVIL ENGINEERS, 345 E. 47th St., New York, N.Y. 10017. Tel.: (212) 752-6800. Divisions: Air Transport, Construction, Engineering Mechanics, Environmental Engineering, Highway, Hydraulics, Irrigation, Pipeline, Power, Soil Mechanics, Structural, Urban Transportation, Surveying, Urban Planning, Waterways, Harbors and Coastal Engineering, Aerospace, Ocean Engineering, and Water Resources Planning and Management. Publications:

Civil Engineering, monthly
Directory

Structural Journal, monthly
Geotechnical Engineering Journal, monthly
Hydraulics Journal, monthly
Engineering Mechanics Journal, bimonthly
Environmental Engineering Journal, bimonthly
Engineering Issues Journal, quarterly
Irrigation and Drainage, quarterly
Waterways, Harbors, and Coastal Engineering, quarterly
Transportation Engineering, quarterly
Construction, aperiodic
Power, aperiodic
Surveying and Mapping, aperiodic
Urban Planning and Development, aperiodic

AMERICAN SOCIETY OF CONSULTING PLANNERS, 1750 Old Meadow Rd., McLean, Virginia 22101. Tel.: (703) 893-7337. Organization of private firms engaged in urban and regional planning. Periodical:
Newsletter, 10 times per year

AMERICAN SOCIETY OF GOLF COURSE ARCHITECTS, 221 N. LaSalle St., Chicago, Illinois 60601. Tel.: (312) 372-7090.

AMERICAN SOCIETY OF HEATING, REFRIGERATING, AND AIR-CONDITIONING ENGINEERS, INC., 345 E. 47th St., New York, N.Y. Tel.: (212) 752-6800. Publication:
ASHRAE Journal, monthly

AMERICAN SOCIETY OF MECHANICAL ENGINEERS, 345 E. 47th St., New York, N.Y. 10017. Tel.: (212) 752-6800. Professional society that developed the boiler and pressure vessel code and is active in the preparation of numerous standards. Publications:
Journal of Applied Mechanics, quarterly
Journal of Dynamic Systems, quarterly
Journal of Engineering for Industry, quarterly
Journal of Engineering for Power, quarterly
Journal of Engineering Materials and Technology, quarterly
Journal of Fluids Engineering, quarterly
Journal of Heat Transfer, quarterly
Journal of Lubrication Technology, quarterly
Journal of Pressure Vessel Technology, quarterly

AMERICAN SOCIETY OF PHOTOGRAMMETRY, 105 N. Virginia Ave., Falls Church, Virginia 22046. Tel.: (703) 534-6617. For individuals and firms that use photogrammetric techniques for city planning, mapping, and other purposes. Publication:
Photogrammetric Engineering, monthly

AMERICAN SOCIETY OF PLUMBING ENGINEERS, 16161 Ventura Blvd., Suite 105, Encino, California 91316. Tel.: (213) 784-4845. Membership includes plumbing engineers, designers, draftsmen, specifiers of plumbing systems, governmental authorities, and others interested in the plumbing engineering profession. Publication:
Plumbing Engineer, monthly

AMERICAN SOCIETY OF SAFETY ENGINEERS, 850 Busse Highway, Park Ridge, Illinois 60068. Tel.: (312) 692-4121. Publication:
Journal of the American Society of Safety Engineers, monthly

AMERICAN SOCIETY OF SANITARY ENGINEERING, 960 Illuminating Bldg., Cleveland, Ohio 44113. Tel.: (216) 696-3228. Conducts research on plumbing and has developed plumbing codes. Publications:
News Letter, monthly
Yearbook

AMERICAN SPECIFICATION INSTITUTE, 134 N. LaSalle St., Chicago, Illinois 60602. Tel.: (312) 236-4214. For persons who write architectural and engineering specifications. Publication:
Specification Record

AMERICAN WELDING SOCIETY, 2501 N.W. 7th St., Miami, Florida 33125. Tel.: (305) 642-7090. Publications:
Welding Journal, monthly
Welding Handbook, annual

ASSOCIATION OF IRON AND STEEL ENGINEERS, Suite 2350, Three Gateway Center, Pittsburgh, Pennsylvania 15222. Tel.: (412) 281-6323.

ASSOCIATION OF WOMEN IN ARCHITECTURE, P.O. Box 1, Clayton, Missouri 63105. Tel.: (314) 721-3909.

BRAB BUILDING RESEARCH INSTITUTE, 2101 Constitution Ave., N.W., Washington, D.C. 20418. Tel.: (202) 961-1515. Encourages and stimulates research and development. Members are architects, engineers, manufacturers, contractors, distributors, building owners, homebuilders, representatives of technical and professional societies, educational and research institutes, and of government. For publications, see BRAB in Section V, GOVERNMENT PLANNING, STANDARDS, CODES, TESTING, INSPECTION.

COLUMN RESEARCH COUNCIL, Fritz Engineering Laboratory, Lehigh University, Bethlehem, Pennsylvania 18015. Tel.: (215) 691-7000, Ext. 452. The council is affiliated with the Engineering Foundation and is devoted to studies and procedures for stability of structures. Publishes guide to design criteria for metal compression members.

CONSTRUCTION SPECIFICATIONS INSTITUTE, Suite 300, 1150 17th St., N.W., Washington, D.C. 20036. Tel.: (202) 833-2160. Composed of individuals who specify design and construction requirements. Concerned with systems and format for making the specifying process more efficient. Publication:

Construction Specifier, monthly

CONSULTING ENGINEERS COUNCIL OF METROPOLITAN WASHINGTON, 8811 Colesville Rd., Suite 225, Silver Spring, Maryland 20910. Tel.: (301) 588-6616. Publications:

CEC/MW Monthly Newsletter
Legislative Alert

COUNCIL OF EDUCATIONAL FACILITY PLANNERS, INTERNATIONAL, 29 W. Woodruff Ave., Columbus, Ohio 43210. Tel.: (614) 422-1521. Organization of individuals involved in planning, designing, creating, equipping, and maintaining the physical environment for education. Publication:

CEFP Journal, bimonthly

EARTHQUAKE ENGINEERING RESEARCH INSTITUTE, 424 40th St., Oakland, California 94609. Tel.: (415) 655-6699. Restricted to persons engaged in design of earthquake-resistant structures, government regulation of earthquake-resistant structures, or studies of earthquakes.

ENGINEERS JOINT COUNCIL, 345 E. 47th St., New York, N.Y. 10017. Tel.: (212) 752-6800. Council of 36 engineering societies that serves as a coordinating agency for engineering activities.

ENVIRONMENTAL ENGINEERING INTERSOCIETY BOARD, INC., Box 9728, Washington, D.C. 20016. Tel.: (202) 232-8553.

FEDERATION OF SOCIETIES FOR PAINT TECHNOLOGY, 121 S. Broad St., Philadelphia, Pennsylvania 19107. Tel.: (215) 545-1506. Publication:
Journal of Paint Technology, monthly

FOOD FACILITIES CONSULTANTS SOCIETY, 600 S. Michigan Ave., Chicago, Illinois 60605. Tel.: (312) 427-2487. Members plan facilities and design equipment for restaurants and institutions.

ILLUMINATING ENGINEERING SOCIETY, 345 E. 47th St., New York, N.Y. 10017. Tel.: (212) 752-6800. Professional society for those engaged in the teaching or design of illumination. Publications:
Journal of the Illuminating Engineering Society, quarterly
Lighting Design and Application, monthly

INSTITUTE OF ELECTRICAL AND ELECTRONICS ENGINEERS, INC., 345 E. 47th St., New York, N.Y. 10017. Tel.: (212) 752-6800. More than 160,000 engineers, scientists, and students involved with electrical engineering and electronics.

INSTITUTE OF STORE PLANNERS, Box 538, Grand Central Station, New York, N.Y. 10017. Tel.: (212) 735-5370.

INTERNATIONAL SOCIETY OF FOOD SERVICE CONSULTANTS, 20 Salt Landing, Tiburon, California 94920. Tel.: (415) 388-9305. Members design food service facilities. Publications:
The Consultant, quarterly
Directory, biennial

NATIONAL ACADEMY OF DESIGN, 1083 5th Ave., New

York, N.Y. 10028. Tel.: (212) 369-4880.

NATIONAL ACADEMY OF ENGINEERING, 2101 Constitution Ave., N.W., Washington, D.C. 20418. Tel.: (202) 961-1658. Publication:
The Bridge, six times per year

NATIONAL ASSOCIATION OF COUNTY ENGINEERS, Linn County Court House, Cedar Rapids, Iowa 52401. Organization of 1,100 engineers employed by counties; aims to advance county road engineering and management. Publication:
NACE Newsletter, quarterly

NATIONAL ASSOCIATION OF CORROSION ENGINEERS, 2400 West Loop South, Houston, Texas 77027. Tel.: (713) 622-8980. Publications:
Corrosion, monthly
Materials Protection and Performance, monthly
Corrosion Abstracts, bimonthly

NATIONAL ASSOCIATION OF GOVERNMENT ENGINEERS, 815 15th St., N.W., Washington, D.C. 20005. Tel.: (202) 638-3195. Publication:
Government Engineers Bulletin, quarterly

NATIONAL ASSOCIATION OF WOMEN IN CONSTRUCTION, 2800 W. Lancaster Ave., Fort Worth, Texas 76107. Tel.: (817) 335-9711. Organization of over 6,000 women employees and employers; aims to improve the skills and knowledge of members and to serve the industry. Publications:
National Newsletter, six times per year
NAWIC Image, quarterly

NATIONAL COUNCIL OF ACOUSTICAL CONSULTANTS, INC., 484 E. Main St., East Aurora, New York 14052. Tel.: (716) 652-0282. Composed of firms that engage in acoustical consulting.

NATIONAL COUNCIL OF ARCHITECTURAL REGISTRATION BOARDS, 2100 M. St., N.W., Suite 706, Washington, D.C. 20037. Tel.: (202) 659-3996. Made up of officials of state architectural registration boards. Publications:

Newsletter, semiannual
Annual Report

NATIONAL COUNCIL OF ENGINEERING EXAMINERS, P.O. Box 752, Clemson, South Carolina 29631. Tel.: (803) 654-2246. Made up of officials of state boards of registration for professional engineers and land surveyors.

NATIONAL COUNCIL ON THE AGING, INC., 1828 L St., N.W., Washington, D.C. 20036. Tel.: (202) 223-6250. Studies problems of aging and develops means for solving them through information, consultation, and research. Publications:
Centers for Older People, three times per year
Current Literature on Aging, quarterly
Directory of National Organizations with Programs on Aging, annual
National Directory of Housing for Older People, annual

NATIONAL SOCIETY OF PROFESSIONAL ENGINEERS, 2029 K St., N.W., Washington, D.C. 20006. Tel.: (202) 337-0420. Organization of more than 70,000 professional engineers in virtually all branches of engineering practice. Publications:
Professional Engineer, monthly
Engineer in Education Newsletter, quarterly
Engineer in Government Newsletter, monthly
Professional Engineer in Industry Newsletter, monthly
Private Practice News, monthly

REFRIGERATING ENGINEERS AND TECHNICIANS ASSOCIATION, INC., 435 N. Michigan Ave., Chicago, Illinois 60611. Tel.: (312) 644-6830. Made up of those who design, install, and maintain industrial air-conditioning and refrigerating equipment.

SOCIETY FOR EXPERIMENTAL STRESS ANALYSIS, 21 Bridge Sq., Westport, Connecticut 06880. Tel.: (203) 227-0829. For engineers who use laboratory techniques to measure stress in structural materials. Publication:
Experimental Mechanics, monthly

REINFORCED CONCRETE RESEARCH COUNCIL, 5420 Old Orchard Rd., Skokie, Illinois 60076. Affiliated with ASCE.

Sponsors research in reinforced concrete.

SOCIETY OF AMERICAN REGISTERED ARCHITECTS, 600 S. Michigan Ave., Chicago, Illinois 60605.
Fosters professionalism in building construction and seeks to make the public more aware of the architect's role. Publication:
Newsletter, ten times per year

SOCIETY OF AMERICAN VALUE ENGINEERS, 2550 Hargrove Dr., L-205, Atlanta, Georgia 30080. Tel.: (404) 436-9508.
Publication:
Journal of Value Engineering, quarterly

SOCIETY OF FIRE PROTECTION ENGINEERS, 60 Batterymarch St., Boston, Massachusetts 02110. Tel.: (607) 482-0686. Publications:
Fire Technology, quarterly
SFPE Bulletin, quarterly

SOCIETY OF MINING ENGINEERS OF AIME, 345 E. 47th St., New York, N.Y. 10017. Publications:
KWIC Index of Rock Mechanics Literature
Rock Mechanics Abstracts Bulletin

Periodicals

ACTUAL SPECIFYING ENGINEER, published monthly by Cahners Publishing Co., Inc., 5 S. Wabash Ave., Chicago, Illinois 60603. For engineers designing and specifying the mechanical and electrical systems and equipment in commercial, institutional, and industrial construction. Covers air-conditioning, heating, plumbing, power generation, electrical distribution, ventilating, refrigeration, lighting, etc. No cost to qualified readers; $10 per year for others.

AIA JOURNAL, published monthly by the American Institute of Architects, 1785 Massachusetts Ave., N.W., Washington, D.C. 20036. For registered architects who design, select, and specify products, and perform other architectural services for buildings and other structures in the United States. No cost to qualified readers; $5 per year for others.

ARCHITECTS' JOURNAL, published weekly by the Architectural Press, 9 Queen Anne's Gate, London SW1, England. Studies of specific buildings with detailed cost analyses; articles on design techniques and problems; examinations of advance working details; guides to the design, construction, and information sheets dealing with building design. Overseas subscription, £12 annually.

ARCHITECTURAL DESIGN, COST & DATA, published monthly by Allan Thompson Publishers, P.O. Box 796, Glendora, California 91740. Costs and data of actual buildings for filing in three-ring binders. Each case study is filed by category: medical, residential, commercial, educational, offices, etc. A preliminary cost guide for architects, contractors, and financial influences over building.

ARCHITECTURAL FORUM, published ten times a year by Whitney Publications, Inc., 130 E. 59th St., New York, N.Y. 10022. For practicing United States registered architects. No cost to qualified readers; $12 per year to others.

ARCHITECTURAL NEWS, published monthly by Gordon Publications, Inc., 20 Community Place, Morristown, New Jersey 07960. For architects and engineers responsible for design and product specification in commercial, industrial, and residential building. News of the profession, legal aspects, government news, new product developments, personnel changes, etc. No cost to qualified readers.

ARCHITECTURAL RECORD, published monthly by McGraw-Hill Publications, 1221 Avenue of the Americas, New York, N.Y. 10020. For architects and engineers engaged in building design and product specification in four primary groups: architects in private practice, consulting engineers, staff architects and engineers in commerce, industry, and institutions, staff architects and engineers in government. $6.60 per year. Special issue: Houses and Apartments.

ARCHITECTURAL REVIEW, published monthly by the Architectural Press, 9 Queen Anne's Gate, London SW1, England. Overseas subscription, £12 annually.

ARCHITECTURE & INTERIORS, published monthly by Architecture & Interiors, 15044 Ventura Blvd., Sherman Oaks, California 91403. For architects, interior designers, builders, and others who specify architectural and interior design products and furniture, furnishings, and accessories. $10 per year.

ARCHITECTURE PLUS, published monthly by Informat Publishing Corp., 1345 Avenue of the Americas, New York, N.Y. 10019. Focuses on changes occurring in architecture and building throughout the world. No cost to qualified readers; $18 per year for others.

ASHRAE JOURNAL, published monthly by American Society of Heating, Refrigerating, and Air-Conditioning Engineers, 345 E. 47th St., New York, N.Y. 10017. For those responsible for design, specification, purchase, installation, operation, and maintenance of heating, refrigerating, air-conditioning, and ventilating systems or components. $10 per year.

BUILDING, published weekly by Building (Publishers) Ltd., The Builder House, P.O. Box 135, 4, Catherine St., London WC2B5JN, England. For architects, quantity surveyors, construction engineers, contractors, and clients. Outside U.K., £9 per year.

BUILDING DESIGN & CONSTRUCTION, published monthly by Industrial Publications, Inc., 5 S. Wabash Ave., Chicago, Illinois 60603. Current events and technology in the commercial, institutional, and industrial building field for architects, engineers, and general contractors. No cost to qualified readers; $10 per year for others. Special issue: Annual BD&C Specifying/Buying Guide and Directory.

BUILDING RESEARCH AND PRACTICE, formerly BUILD INTERNATIONAL, published bimonthly by Leyden Publishing Co. Ltd., 5-7 Carnaby St., London W1A4XT, England. Forecasts trends and provides data and tables on current research of interest to architects, contractors, engineers, materials and components manufacturers, system builders and designers, etc. £8 per year.

BUILT ENVIRONMENT, published monthly by Building (Publishers) Ltd., The Builder House, P.O. Box 135, 4, Catherine St., London WC2B5JN, England. For architects, planners, and engineers. Outside U.K., £6.50 per year.

CHARETTE, published monthly by Archimedia, Inc., 100 N. Lambert St., Philadelphia, Pennsylvania 19103. Problems, legalities, and complexities of architectural practice. No cost to qualified members; $10 per year for others.

CIVIL ENGINEERING, published monthly by American Society of Civil Engineers, 345 E. 47th St., New York, N.Y. 10017. For the engineered construction market—highways, bridges, water supply and waste treatment facilities, urban development, etc. $7 per year.

COMMERCE BUSINESS DAILY, Architectural Engineering Services, published daily, Monday through Friday, by the Government Printing Office, Washington, D.C. 20402. Government listing of Federal projects for which engineers and architects will be retained. $25 per year.

COMMODITY FILE, published by Building (Publishers) Ltd., The Builder House, P.O. Box 135, 4, Catherine St., London WC2B 5JN, England. Reference and product information in six loose-leaf binders, updated each week in a special subscribers' edition of *Building*. Outside U.K., first year £45, thereafter £25 per year.

CONSULTING ENGINEER, Published monthly by Consulting Engineer Publishing Co., 217 Wayne St., St. Joseph, Michigan 49085. For consulting engineers in private practice—partners, principals, department heads, project leaders, and other engineers who have responsibility for specifications and for the engineering services performed for their clients. No cost to qualified readers; $10 per year for others.

DESIGN & ENVIRONMENT, published quarterly by RC Publications, Inc., 19 W. 44th St., New York, N.Y. 10036. Reports on environmental design as approached by the architect, engineer, industrial designer, interior designer, graphic designer, landscape architect, and the city planner. $10 per year.

F. W. DODGE CONSTRUCTION NEWS, published in four editions by McGraw-Hill Information Systems Co., 230 W. Monroe St., Chicago, Illinois 60606. Serves the annual nation conventions of the American Institute of Architects and the Construction Specifications Institute. Provides a complete chronology of each convention plus news of interest to the professions. New products and systems are included. No cost to qualified readers.

ELECTRICAL CONSULTANT, published monthly by W. R. C. Smith Publishing Co., 1760 Peachtree Rd., N.W., Atlanta, Georgia 30309. For profession engineers who design and write specifications for electrical and lighting systems in office buildings, institutions, schools, factories, apartments, and such special projects as airports, expositions, and stadiums. No cost to qualified readers; $15 per year for others.

ELECTRICAL WORLD, published semimonthly by McGraw-Hill Publications, 1221 Avenue of the Americas, New York, N.Y. 10020. Innovations in planning, design, construction, operation, and maintenance of facilities for generation, transmission, distribution, and coordinated control of interconnected electrical systems. $10 per year.

ELECTRICAL LIGHT & POWER, ENERGY/GENERATION EDITION, published monthly by Cahners Publishing Co., 221 Columbus Ave., Boston, Massachusetts 02116. Interpretive, technical, and management information for planning, design, funding, construction, operation, and maintenance of power generating systems that convert fuel energy to electric energy. No cost to qualified readers; $15 per year for others.

ELECTRIC LIGHT & POWER, TRANSMISSION/DISTRIBUTION EDITION, published monthly by Cahners Publishing Co., 221 Columbus Ave., Boston, Massachusetts 02116. Interpretive, technical, and management information for economic management, planning, design, procurement, construction, operation, and maintenance of electric utility transmission and distribution systems. No cost to qualified readers; $15 per year for others.

ELECTRIC UTILITY GENERATION PLANBOOK, published annually by McGraw-Hill Publications, 1221 Avenue of the Americas, New York, N.Y. 10020. Current trends and techniques in fuels, plant siting, steam system design and equipment as related to the electric utility in one reference volume. No cost to qualified readers.

EMPIRE STATE ARCHITECT, official publication of New York State Association of Architects. Published quarterly by Harry Gluckman Co., 126 S. Elmwood Ave., Buffalo, N.Y. 14202. News, technical, and educational matter related to architecture and the practice of architecture within the state. $10 per year.

FLORIDA ARCHITECT, published bimonthly by the Florida Association of the American Institute of Architects, 7100 N. Kendall Dr., Miami, Florida 33156. No cost to qualified readers.

HEATING/PIPING/AIR-CONDITIONING, published monthly by Reinhold Publishing Co., 10 S. La Salle St., Chicago, Illinois 60603. Design, installation, operation, maintenance, expansion, modernization, and replacement techniques as applied to heating and air-conditioning systems and to industrial, process, and power piping systems. $7 per year. Special issue: HPAC Info-dex Mechanical Systems Information Index.

HOUSE & HOME, published monthly by McGraw-Hill Publications, 1221 Avenue of the Americas, New York, N.Y. 10020. For those whose skills in planning and architecture, building methods and technology, land buying and development, finance, management, and marketing are utilized in housing and other light construction markets. $6 per year.

INLAND ARCHITECT, official publication of the Illinois Council and Chicago Chapter, American Institute of Architects. Published 11 times a year by Inland Architect Corp., 1800 S. Prairie, Chicago, Illinois 60616. Regional professional journal edited for architects, structural engineers, and urban planners in the midwest with emphasis on Chicago and Illinois. No cost to qualified readers; $10 per year for others.

JOURNAL OF THE FLORIDA ENGINEERING SOCIETY, published monthly by Florida Engineering Society, 1906 Lee Rd.,

Orlando, Florida 32810. For those in the engineering profession, including engineers working in private practice, industry, education, and municipal, county, state, and Federal government. $5 per year.

MILITARY ENGINEER, published bimonthly by Society of American Military Engineers, 800 17th St., N.W., Washington, D.C. 20006. Covers projects being conducted by the engineering services of the Armed Forces and by industry and by civilian engineers in professional practice, including construction of buildings, roads, bridges, airfields, missile bases, and river and harbor work; military engineer operations, training, and expedients; surveying and mapping; heavy equipment; electronics; space developments; natural resources and materials. $12.50 per year.

MONTHLY BULLETIN OF THE MICHIGAN SOCIETY OF ARCHITECTS, published monthly by the Michigan Architectural Foundation, 28 W. Adams St., Detroit, Michigan 48226. Provides architects and engineers in the area with current information about the building industry and current technology and research on new materials and products and their applications. $4 per year. Special issues: Membership Directory (January); Architectural Firm Roster.

NORTHWEST ARCHITECT, published seven times a year by the Bruce Publishing Co. for the Minnesota Society of Architects, 1120 Glenwood Ave., Minneapolis, Minnesota 55405. For architects, engineers, specifications writers, and construction executives in Minnesota, North Dakota, South Dakota, Wisconsin, Wyoming, Iowa, and eastern Montana. No cost to qualified readers; $5 per year for others.

OHIO ENGINEER, published monthly by the Ohio Society of Professional Engineers, 445 King Ave., Columbus, Ohio 43201. News articles on activities of the society, public relations programs, achievements of professional engineers, the professional engineer's role in community activities, legislation affecting engineers, etc. $5 per year.

PENNSYLVANIA PROFESSIONAL ENGINEER, official publication of the Pennsylvania Society of Professional Engineers. Published bimonthly by Initia, Inc., Old Post Office Bldg., Allegheny Square West, Pittsburgh, Pennsylvania 15212. Reports on construction, specifications, and problems indigenous to the state. $3 per year. Directory issued annually.

POWER ENGINEERING, published monthly by Technical Publishing Co., 1301 S. Grove Ave., Barrington, Illinois 60010. Readers are responsible for design, specifications, operation, and maintenance of power systems and equipment used to generate, transmit, distribute, and utilize power. No cost to qualified readers; $10 per year for others.

PROFESSIONAL ENGINEER, published monthly by National Society of Professional Engineers, 2029 K St., N.W., Washington, D.C. 20006. For professional engineers in consulting, contracting, industry, government, and research. Specialties include civil, sanitary, mechanical, electrical, and other engineering disciplines. $7 per year.

PROGRESSIVE ARCHITECTURE, published monthly by Reinhold Publishing Co., 600 Summer St., Stamford, Connecticut 06904. Feature articles discuss new buildings and contemplated projects; trends in design and construction; social, political, economic forces affecting building design; philosophy of building design; advances in building technology; developments in interior design. Regular columns discuss specifications writing; mechanical and electrical engineering; legal aspects of architectural practice. $5 per year.

SYMPOSIA, published monthly by Boyce Publications, Inc., 4076 Estes St., Wheat Ridge, Colorado 80033. For the architectural/construction community in the Rocky Mountain region. Concerned with architectural, engineering, and mechanical design studies, specification writing, construction organization activity, legislative matters affecting the industry, management efficiency, bidding procedures, ethical practices, etc.

TEXAS PROFESSIONAL ENGINEER, published monthly by the Texas Society of Professional Engineers, P.O. Box 2145, Austin, Texas 78767. For members of the society, including engineers representing industry, consultants, municipal government, government agencies, colleges and universities. $3 per year; Directory Issue, $5.

TRANSMISSION & DISTRIBUTION, published monthly by Cleworth Publishing Co., 1 River Rd., Cos Cob, Connecticut 06807. Covers design, construction, operation, maintenance, purchasing for transmission and distribution lines, substations, outdoor lighting, metering, communications, and control. No cost to qualified readers; $12 per year for others.

UNDERGROUND ENGINEERING, published bimonthly by Technology Publishing Corp., 825 S. Barrington Ave., Los Angeles, California 90049. For utility managers, underground design and project engineers, supervisors, and engineering group managers responsible for the research and development, design, production, installation, testing, evaluation, purchase, and operation of subterranean power systems, products, and processes. $20 per year.

UNDERGROUNDING, published bimonthly by Asman Corp., 15300 Ventura Blvd., Suite 220, Sherman Oaks, California 91403. For those in the fields of electrical power, telephone, and cable television responsible for underground engineering. Reports on formation and development of cooperative ventures, development of legislation and public policy; development and usage of new materials, equipment, processes, systems, and techniques. No cost to qualified readers; $15 per year for others.

WESTERN BUILDING DESIGN, published monthly by American Institute of Building Design, 1830 W. 8th St., Suite 305, Los Angeles, California 90057. For licensed architects, building designers, and specifiers of products in the West who are engaged in the many types of design—single family, multiple family, remodeling, schools, churches, commercial/industrial, and public works. No cost to qualified readers; $6 per year for others.

WISCONSIN PROFESSIONAL ENGINEER, published monthly from September to June by the Wisconsin Society of Professional Engineers, 1618 Beltline Highway, Madison, Wisconsin 53713. $2 per year for members; $3 per year for nonmembers. Directory published annually.

Directories

ASCE DIRECTORY. Listing of all members, their geographical distribution, and a historic roster of all honorary members. American Society of Civil Engineers, 345 E. 47th St., New York, N.Y. 10017. Available only to ASCE members: $5.

AMERICAN ARCHITECTS DIRECTORY. Biographical information on over 20,000 architects, including professional affiliation, address, education, etc. R. R. Bowker Co., 1180 Avenue of the Americas, New York, N.Y. 10036. $35.

AACE DIRECTORY OF MEMBERS. American Association of Cost Engineers, 308 Monongahela Bldg., Morgantown, West Virginia 26505. Free to members; not available to nonmembers.

AMERICAN ASSOCIATION OF HOSPITAL CONSULTANTS DIRECTORY. Listing of consultants and their specialties. American Association of Hospital Consultants, 1700 K St., N.W., Washington, D.C. 20006.

AIA MEMBERSHIP DIRECTORY. Listing of all AIA members with addresses and chapter affiliations. Includes such information as component presidents, headquarters staff, affiliated organizations, trade and professional associations, foreign societies, fellows, and architectural schools. American Institute of Architects, 1735 New York Ave., N.W., Washington, D.C. 20006. AIA members, $5; others, $25.

AMERICAN MANAGEMENT ASSOCIATION DIRECTORY OF CONSULTANT MEMBERS. List of organizations and individual consultants who are members of AMA, and their services. American Management Association, 135 W. 50th St., New York, N.Y. 10020. AMA members, $4; others, $5.25.

ARCHITECTURAL INDEX. Using nine major U.S. periodicals on architecture and land planning, the editor indexes information on building types, architecture, materials, and construction techniques. Erwin J. Bell, P.O. Box 1168, Boulder, Colorado 80302. $7.

ARCHITECTURAL PERIODICALS INDEX. Subject fields from over 300 international publications include architecture and allied arts, construction technology, design, environmental studies, planning, and relevant research. RIBA Publications Ltd., 66 Portland Place, London W1N 4AD, England. £10.

ARCHITECTURAL PRESS INTERNATIONAL DIRECTORY OF COMPUTER PROGRAMS FOR THE CONSTRUCTION INDUSTRY. Abstracts of 900 programs relevant to the construction industry, with details of what they do and where to obtain them. Topics include management, quantity surveying, mechanical, structural, and H&V design and graphics, and building layout. The Architectural Press, 9 Queen Anne's Gate, London SW1, England. £12.

ASHRAE HANDBOOK AND PRODUCT DIRECTORY. Complete directory of manufacturers' names and addresses, complete directory of products in the field, and trade name index. American Society of Heating, Refrigerating, and Air-Conditioning Engineers, 345 E. 47th St., New York, N.Y. 10017.

ASSOCIATION OF CONSULTING MANAGEMENT ENGINEERS DIRECTORY OF MEMBERSHIP AND SERVICES. Profile of services available from each member-consultant. Association of Consulting Management Engineers, 347 Madison Ave., New York, N.Y. 10017. Free.

ASSOCIATION OF UNIVERSITY ARCHITECTS MEMBERSHIP ROSTER. Association of University Architects, Western Washington State College, Bellingham, Washington 98225. No charge.

ASSOCIATION OF WOMEN IN ARCHITECTURE MEMBERSHIP DIRECTORY. Biographical information on members. Association of Women in Architecture, P.O. Box 1, Clayton, Missouri 63105. $1.

BIENNIAL REPORT AND ROSTER OF REGISTERED PROFESSIONAL ENGINEERS. Names and addresses of registered professional engineers in New Hampshire. New Hampshire Board of Registration for Professional Engineers, c/o Secretary of State, State House, Concord, New Hampshire 03301. $1.

BUILDING CONSTRUCTION INFORMATION SOURCES,

H. B. Bentley, ed. How to get information on nearly 1000 building topics. Gale Research Co., Book Tower, Detroit, Michigan 48226. $14.50.

CENTER FOR BUILDING TECHNOLOGY LIST OF PUBLICATIONS. Publications, authors, dates, and cost on such subjects as building systems, codes and standards, concrete, environment, fire research, housing, building materials, plumbing, roofs, and structures. U.S. Dept. of Commerce, National Bureau of Standards, Center for Building Technology, Washington, D.C. 20234. Order No. LP57; no charge.

COLUMBIA UNIVERSITY CATALOG OF THE AVERY MEMORIAL ARCHITECTURAL LIBRARY. One of the chief scholarly tools available to the profession. New acquisitions emphasize literature on urban renewal and the social aspect of urban design. G.K. Hall and Co., 70 Lincoln St., Boston, Massachusetts 02111. 1968, 19 volumes, 15,883 pp., $1,140. First Supplement, 1973, 4 volumes, 3,166 pp., $390.

COLUMBIA UNIVERSITY AVERY INDEX TO ARCHITECTURAL PERIODICALS. Comprehensive index including decorative arts, archaeology, and physical aspects of city planning and housing. 362,000 entries. G.K. Hall and Co., 70 Lincoln St., Boston, Massachusetts 02111. 1973, 15 volumes, $1,235.

CONSULTANTS AND CONSULTING ORGANIZATIONS DIRECTORY, P. Wasserman and J. McLean, eds. Reference guide to concerns and individuals engaged in consultation for business, industry, and government. 5041 entries, 146 subject fields. Gale Research Co., Book Tower, Detroit, Michigan 48226. $45.

CONSULTING ENGINEERS COUNCIL/USA. List of U.S. consulting engineers and consulting companies, Consulting Engineers Council/USA, 1155 15th St., N.W., Suite 713, Washington, D.C. 20005. $10.

DICTIONARY CATALOG OF THE UNITED STATES DEPARTMENT OF HOUSING AND URBAN DEVELOPMENT

LIBRARY AND INFORMATION DIVISION. Catalog of urban and regional data, much of which has been compiled through HUD-sponsored studies. Library contains over 12,000 Comprehensive Planning Reports and Model Cities reports, plus holdings of the libraries of the Federal Housing Administration, Public Housing Administration, and the Housing and Home Finance Agency. G.K. Hall and Co., 70 Lincoln St., Boston, Massachusetts 02111. 1973, 19 volumes, $1,425.

DIRECTORY OF ENGINEERS REGISTERED IN NORTH CAROLINA. Industrial Extension Service, North Carolina State University, Raleigh, North Carolina 27607. $5.

ENCYCLOPEDIA OF ASSOCIATIONS. *Vol. 1, National Organizations of the U.S.* 17,866 entries. Covers nonprofit associations, societies, federations, unions, etc., in 18 major categories: agriculture, business, education, etc. 1,546 pp., $45. *Vol. 2, Geographic and Executive Index.* Geographic section lists groups in Vol. 1 by state and city, with phone numbers and addresses. Executive section lists the name of the chief officer of each organization. 512 pp. $28.50. *Vol. 3, New Associations and Projects,* quarterly supplement service. Interedition subscription, with binders for new subscribers, $48. Gale Research Co., Book Tower, Detroit, Michigan 48226.

ENGINEERING CONSULTANTS—AICE MEMBERS. List of members with biographical and specialty information. American Institute of Consulting Engineers, 345 E. 47th St., New York, N.Y. 10017. $3.50.

ENGINEERING INDEX MONTHLY. More than 3,500 journals, proceedings, reports, etc., are reviewed each year for engineering significance and abstracted monthly. Engineering Index Monthly, 345 E. 47th St., New York, N.Y. 10017. $400 annually.

ENGINEERING SOCIETIES LIBRARY CLASSED SUBJECT CATALOG. Catalog for the largest engineering library in the U.S.—an archive for older material and many special collections and a source of current information. G.K. Hall and Co., 70 Lincoln St., Boston, Massachusetts 02111. 1963, 12 volumes, $930. Supplements for later years.

ENGINEERS JOINT COUNCIL DIRECTORY OF ENGINEERING SOCIETIES AND RELATED ORGANIZATIONS. Data on more than 300 national, regional, and international organizations concerned with engineering. Engineers Joint Council, 345 E. 47th St., New York, N.Y. 10017. Order No. 101-70, $8.

ENGINEERS OF DISTINCTION—A WHO'S WHO IN ENGINEERING. 5,000 eminent engineers from all fields—industry, education, government, and private practice. Contains biographies, tabulates and describes national awards by societies, with recipients. Geographical and specialty index. Engineers Joint Council, 345 E. 47th St., New York, N.Y. 10017. Order No. 107-73, $35.

GUIDE TO ARCHITECTURAL INFORMATION. Sourcebook for major library references for information on architecture, housing, and urban design. Design Data Center, P.O. Box 566, Lansdale, Pennsylvania 19446. $4.95.

HARVARD UNIVERSITY CATALOG OF THE LIBRARY OF THE GRADUATE SCHOOL OF DESIGN. Collection includes material on urban planning in such areas as urban renewal, urban design, city and regional planning, state and national planning, housing, zoning, and regional science. G.K. Hall and Co., 70 Lincoln St., Boston, Massachusetts 02111. 1968, 44 volumes, 28,327 pp., $2,860. Supplement for later years.

HOW TO FIND OUT IN ARCHITECTURE AND BUILDING—A GUIDE TO SOURCES OF INFORMATION, D.L. Smith. Sources of information on architecture, planning, and building. Pergamon Press, Maxwell House, Fairview Park, Elmsford, New York 10523.

LEARNING RESOURCES. Directory for engineers, scientists, managers, and educators. Descriptive information about courses, seminars, conference, workshops, and other continuing education activities in the U.S. and Canada. Updated and issued three times a year. Engineers Joint Council, 345 E. 47th St., New York, N.Y. 10017. $24.

NATIONAL RESEARCH COUNCIL OF CANADA LIST OF

PUBLICATIONS. 1947-71, 1972, and 1973. Lists of papers on such subjects as building research, Canadian Building Digests, fire studies, Canadian Building Abstracts, Housing Notes, computer programs, and the National Building Code of Canada. National Research Council of Canada, Division of Building Research, Ottawa, Canada.

NATIONAL SOCIETY OF PROFESSIONAL ENGINEERS DIRECTORY OF ENGINEERS IN PRIVATE PRACTICE. Names, addresses, and phone numbers of more than 3,000 registered professional engineers. Services offered by 1,200 consulting engineering firms are given along with the names of principals, firm addresses, and phone numbers. More than 13,200 entries of firms in ten major engineering specialties are listed. National Society of Professional Engineers, 2029 K St., N.W., Washington, D.C. 20006. Order No. 1919. NSPE members, $6; others, $12.

OHIO ARCHITECT—ROSTER ISSUE. List of registered architects in Ohio. The Architects Society of Ohio, 37 W. Broad St., Columbus, Ohio 42215. $4.

PENNSYLVANIA SOCIETY OF ARCHITECTS DIRECTORY. List of AIA members in Pennsylvania. Chatham Associates, Chamber of Commerce Bldg., Pittsburgh, Pennsylvania.

REGISTERED ARCHITECTS—NEW YORK STATE. List of registered architects in New York. University of the State of New York, State Education Bldg., Albany, New York 12201.

ROSTER OF PROFESSIONAL ENGINEERS AND LAND SURVEYORS. Directory of licensed engineers in the state of Washington. State Board of Registration for Professional Engineers and Land Surveyors, Highways-Licensing Bldg., Olympia, Washington. No charge.

ROSTER OF PROFESSIONAL ENGINEERS AND SURVEYORS. Directory of licensed engineers in California. Board of Registration for Professional Engineers, Rm. A-102, 1021 O St., Sacramento, California 95814. $1.

ROSTER OF REGISTERED ARCHITECTS—FLORIDA. List of registered architects in Florida. Florida State Board of Architecture, P.O. Box 2185, Ellinor Village Station, Ormond Beach, Florida 32074. $2.

SPECIFICATION, D. Harrison, ed. Source of information on building materials and techniques in England. Architectural Press, 9 Queen Anne's Gate, London SW1, England.

TEXAS PROFESSIONAL ENGINEER. Directory of members of the Texas Society of Professional Engineers. Thomas A. Melody, 503 San Jacinto, P.O. Box 2145, Austin, Texas 78767.

WHO'S WHO IN CONSULTING, P. Wasserman et al. Biographical details about 10,000 individuals who work as full-time, part-time, or occasional consultants. Many are on the staffs of private consulting organizations. Others are members of the academic community who accept consulting assignments in their fields. Gale Research Co., Book Tower, Detroit, Michigan 48226. $45.

III. BUILDING MATERIALS AND MANUFACTURERS

In this section are found information sources for the basis of the construction art—from asbestos to zinc, through lumber and steel, materials are the *sine qua non* of construction. Here, the reader can find trade associations eager to promote use of their products by sharing technical know-how. Here, too, the reader will find the "bibles" of the materials industry to help keep up with the latest developments, ideas, and applications.

BUILDINGS AND BUILDING COMPONENTS—INCLUDING HARDWARE, CONSTRUCTION EQUIPMENT

Associations and Professional Groups

ACCESS FLOOR MANUFACTURERS ASSOCIATION, 724 York Rd., Baltimore, Maryland 21204. Tel.: (301) 821-5560. Manufacturers of removable modular floor systems.

ACOUSTICAL DOOR INSTITUTE, U.S. Plywood Co., 1001 Perry St., Algoma, Wisconsin 54201. Tel.: (414) 487-5221. Made up of ten manufacturers of acoustical doors.

AMERICAN HARDBOARD ASSOCIATION, 20 N. Wacker Dr., Chicago, Illinois 60606. Tel.: (312) 236-8009. Represents over 90 percent of the total U.S. production capacity. Provides membership and the public with educational activities and statistical, legislative, and technical information on U.S. hardboard products.

AMERICAN INSTITUTE OF KITCHEN DEALERS, 114 Main St., Hackettstown, New Jersey 07840. Tel.: (201) 852-0033.

AMERICAN LADDER INSTITUTE, 111 E. Wacker Dr., Chicago, Illinois 60601. Tel.: (312) 644-6610. Organization through which wood and metal ladder manufacturers have sponsored ANSI ladder codes. Publication:
Newsletter, monthly

AMERICAN SOCIETY OF ARCHITECTURAL HARDWARE CONSULTANTS, Box 3476, San Rafael, California 94902. Tel.: (415) 479-6340. Group of specialists which has prepared specifications and standards for architectural hardware. Publication:
News and Views, bimonthly

ARCHITECTURAL ALUMINUM MANUFACTURERS ASSOCIATION, 410 N. Michigan Ave., Chicago, Illinois 60611. Tel.: (312) 828-9637. Manufacturers of aluminum storm windows and doors, store fronts, and curtain walls. Publications:
Architectural Aluminum Journal, quarterly
Certified Products Directory, quarterly

ARCHITECTURAL WOODWORK INSTITUTE, 5055 S. Chesterfield Rd., Arlington, Virginia 22206. Tel.: (703) 671-9100. Manufacturers of architectural woodwork.

ASSOCIATED EQUIPMENT DISTRIBUTORS, 615 W. 22nd St., Oak Brook, Illinois 60521. Tel.: (312) 654-0650. Manufacturers and distributors of construction equipment. Publications:
Construction Equipment Distribution
Contact, biweekly

BITUMINOUS EQUIPMENT MANUFACTURERS BUREAU, Suite 1700, Marine Plaza, 111 E. Wisconsin Ave., Milwaukee, Wisconsin 53202. Tel.: (414) 272-0943. Manufacturers of bituminous mixing and paving equipment.

BUILDERS HARDWARE MANUFACTURERS ASSOCIATION, 60 E. 42nd St., New York, N.Y. 10017. Tel.: (212) 682-8142. Group of manufacturers of builders' hardware.

CHURCH FURNITURE MANUFACTURERS ASSOCIATION, 666 N. Lake Shore Dr., Chicago, Illinois 60611. Tel.: (312) 944-5566. Division of the National Association of Furniture Manufacturers.

CONSTRUCTION INDUSTRY MANUFACTURERS ASSOCIATION, 1700 Marine Plaza, 111 E. Wisconsin Ave., Milwaukee, Wisconsin 53202. Tel.: (414) 272-0943. Group of manufacturers of construction equipment.

CONVEYOR EQUIPMENT MANUFACTURERS ASSOCIATION, 1000 Vermont Ave., N.W., Washington, D.C. 20005. Tel.: (202) 628-4634. Publication:
Bulletin, monthly

COUNCIL OF HOUSING PRODUCERS, 1801 Avenue of the Stars, Los Angeles, California 90067. Tel.: (213) 277-8375. Association of 14 housing producers.

CRANE MANUFACTURERS ASSOCIATION OF AMERICA, 1326 Freeport Rd., Pittsburgh, Pennsylvania 15238. Tel.: (412) 782-1624.

FIR AND HEMLOCK DOOR ASSOCIATION, Yeon Bldg., Portland, Oregon 97204. Tel.: (503) 224-3930. Seven manufacturers of fir and hemlock doors. Publication:
Fir and Hemlock Door Specialists, annual

GYPSUM ASSOCIATION, 201 N. Wells St., Chicago, Illinois 60606 and 1800 N. Highland Ave., Hollywood, California 90028. Promotes gypsum products on behalf of member companies, and serves as an information center for the building industry.

HOLLOW METAL DOOR AND BUCK ASSOCIATION, 405 Lexington Ave., Suite 5510, New York, N.Y. 10017. Tel.: (212) 687-2380. Group of manufacturers of steel doors.

INCINERATOR INSTITUTE OF AMERICA, 2425 Wilson Blvd., Arlington, Virginia 22201. Group of manufacturers of incinerators for municipal, commercial, and home use which has developed standards and guides.

MANUFACTURED HOUSING ASSOCIATION OF AMERICA, 39 S. LaSalle St., Chicago, Illinois 60603. Tel.: (312) 263-1748. Association of mobile housing manufacturers and dealers, mobile park operators, and others. Publication:
Monday Morning Hotline, weekly

MATERIAL HANDLING INSTITUTE, 1326 Freeport Rd., Pittsburgh, Pennsylvania 15238. Tel.: (412) 782-1624.

METAL BUILDING MANUFACTURERS ASSOCIATION, Keith Bldg., Cleveland, Ohio 44115. Tel.: (216) 241-7333.

METAL LADDER MANUFACTURERS ASSOCIATION, P.O. Box 580, Greenville, Pennsylvania 16125. Tel.: (412) 588-8600.

MOBILE HOME MANUFACTURERS ASSOCIATION, Box 201, 14650 Lee Rd., Chantilly, Virginia 22021. Group of manufacturers and suppliers of equipment. Promotes mobile home standards. Publication:
Mobile Home Life, annual

NATIONAL ASSOCIATION OF ARCHITECTURAL METAL MANUFACTURERS, 1033 South Blvd., Oak Park, Illinois 60302. Tel.: (312) 383-7725. Manufacturers of metal doors, stairs, curtain walls, and other ornamental and architectural components. Publication:
Architectural Metals, bimonthly

NATIONAL ASSOCIATION OF BUILDING MANUFACTURERS, 1619 Massachusetts Ave., N.W., Washington, D.C. 20036. Tel.: (202) 234-1374. Manufacturers of modular, industrialized buildings. Publications:
Housing and Building, monthly
Technics, quarterly
Builders' Guide to Manufactured Homes, annual

NATIONAL ASSOCIATION OF STORE FIXTURE MANUFACTURERS, 53 W. Jackson Blvd., Chicago, Illinois 60604. Tel.: (312) 922-0509. Publication:
NASFM News, monthly

NATIONAL BUILDERS HARDWARE ASSOCIATION, 1815 N. Ft. Myer Dr., Suite 412, Rosslyn, Virginia 22209. Group of wholesalers of building hardware products. Promotes standards for wood and metal door hardware. Publications:
Doors and Hardware, monthly
NBHA Reporter, bimonthly

NATIONAL ELEVATOR INDUSTRY, INC., 101 Park Ave., New York, N.Y. 10017. Tel.: (212) 532-2327. Group of manufacturers of elevators and related equipment. Publishes elevator installation manual. Publication:
NEII Newsletter, monthly

NATIONAL HOME IMPROVEMENT COUNCIL, 11 E. 44th St., New York, N.Y. 10017. Group of manufacturers, contractors, and lending institutions which fosters interest in home improvements. Publication:
NHIC Newsletter, monthly

NATIONAL HOUSING CONFERENCE, 1250 Connecticut Ave., N.W., Washington, D.C. 20036. Tel.: (202) 223-4844. Organization of individuals, planning officials, and representatives of labor, social, religious, and other groups that are interested in housing and urban renewal. Publications:
Newsletter, monthly
Housing Yearbook

NATIONAL HOUSING PRODUCERS ASSOCIATION, 900 Peachtree St., N.E., Atlanta, Georgia 30309. Tel.: (404) 875-0781.

NATIONAL KITCHEN CABINET ASSOCIATION, 334 E. Broadway, Louisville, Kentucky 40202. Tel.: (502) 583-4328.

NATIONAL LOCKSMITH SUPPLIERS ASSOCIATION, 95 E. Valley Stream Blvd., Valley Stream, New York 11580. Tel.: (516) 825-6673.

NATIONAL PARTICLEBOARD ASSOCIATION, 2306 Perkins Place, Silver Spring, Maryland 20910. Tel.: (301) 587-2204. Group of manufacturers of particleboard which has developed product standards.

NATIONAL SASH AND DOOR JOBBERS ASSOCIATION, 20 N. Wacker Dr., Chicago, Illinois 60606. Tel.: (312) 263-2670. Group of wholesale distributors of sash and doors. Offers millwork home study course. Publications:
Newsletter, monthly
NSDJA Digest, bimonthly

PRODUCERS COUNCIL, 1717 Massachusetts Ave., N.W., Washington, D.C. 20036. Tel.: (202) 667-8727. Represents 130 manufacturers at the national level and over 4,000 at the local level. Interfaces with governmental agencies and specifying and buying influences in construction. Supplies information to potential users on a broad spectrum of building products.

SCAFFOLDING AND SHORING INSTITUTE, 2130 Keith Bldg., Cleveland, Ohio 44115. Tel.: (216) 241-7333.

SCREEN MANUFACTURERS ASSOCIATION, 110 N. Wacker Dr., Chicago, Illinois 60606. Tel.: (312) 346-6191. Group of manufacturers of window and door screens for insect control.

STEEL DOOR INSTITUTE, 2130 Keith Bldg., Cleveland, Ohio 44115. Tel.: (216) 241-7333. Group of manufacturers of metal doors.

STEEL JOIST INSTITUTE, 2001 Jefferson Davis Highway, Arlington, Virginia 22202. Tel.: (703) 920-1700. Group of manufacturers of open web steel joists which has developed standards for design and use of the group's products.

STEEL PLATE FABRICATORS ASSOCIATION, 15 Spinning Wheel Rd., Hinsdale, Illinois 60521. Tel.: (312) 654-3640.

STEEL WINDOW INSTITUTE, 2130 Keith Bldg., Cleveland, Ohio 44115. Tel.: (216) 241-7333. Group of manufacturers of steel windows and window operators.

Periodicals

ARCHITECTURAL METALS, published bimonthly by the National Association of Architectural Metal Manufacturers, 1010 W. Lake St., Oak Park, Illinois 60301. Reports on new developments and new ways of using architectural metals.

AUTOMATION IN HOUSING, published ten times per year by Vance Publishing Corp., 300 W. Adams, Chicago, Illinois 60606. For builders of housing (single and multifamily), mobile homes, and industrial and commercial buildings who perform all or part of the operation under-roof, instead of on-site construction. No cost for qualified readers.

BUILDING PRODUCTS GUIDES, comprising *Home Building & Remodeling* (Jan.), *Vacation Homes & Leisure Living* (Feb.), *Home Planning & Decorating* (Mar.), *Kitchens, Baths, & Family Rooms* (Apr.), *Home Building & Remodeling* (Jul.), *Vacation Homes & Leisure Living* (Aug.), *Home Planning & Decorating* (Sept.), *Kitchens, Baths, & Family Rooms* (Oct.). Published by Hudson Publishing Co., 175 S. San Antonio Rd., Los Altos, California 94022. Reports on new and redesigned products and materials and on new design ideas. $6.95 per year; single copy $1.25.

COMMODITY FILE, published by Building (Publishers) Ltd., The Builder House, P.O. Box 135, 4, Catherine St., London WC2B 5JN, England. Reference and product information in six loose-leaf binders, updated each week in a special subscribers' edition of *Building*. Outside U.K., first year £45; thereafter, £25 per year.

DRYWALL MAGAZINE, published bimonthly by Gypsum Drywall Contractors International, 2010 Massachusetts Ave., N.W., Suite 600, Washington, D.C. 20036. Articles on use of material and systems, construction techniques, general industry news, trends, statistics, personnel changes, law, accounting, legislation, merchandising, costs, management, financing, and government regulations. No cost to qualified readers; $6 per year for others.

ELEVATOR WORLD, published monthly by Elevator World, P.O. Box 6523, Loop Branch, Mobile, Alabama 36606. No cost to qualified readers; $8 per year for others. Special issue: Elevator World Annual Issue (Oct.).

HARDWARE MERCHANDISER, published monthly by Irving-Cloud Publishing Co., 7300 N. Cicero Ave., Lincolnwood, Chicago, Illinois 60646. No cost to qualified readers; $5 per year for others.

KITCHEN & BATH, published bimonthly by Johns Publications Inc., P.O. Box 1744, Palm Springs, California 92262. For kitchen specialists, cabinet manufacturers, home improvement contractors, architects, builders, countertop and vanitory fabricators, appliance dealers, designers, plumbing and heating contractors, utility companies, building supply, hardware, and lighting dealers, and flooring and tile contractors. No cost to qualified readers.

KITCHEN BUSINESS, published monthly by Gralla Publications, 1501 Broadway, New York, N.Y. 10036. For dealers and distributors who sell residential kitchen equipment to the remodeling and new-home markets, and also for kitchen cabinet, countertop, and vanity factories and shops. $2 per year. Annual directory.

METAL BUILDING REVIEW, published monthly at P.O. Box 1255, South Bend, Indiana 46624. For the metal building industry, particularly metal building dealer/builders, and sales and merchandising personnel of manufacturers of metal buildings, their suppliers, and the manufacturers of collateral building materials. No cost to qualified readers; $4 per year for others.

SYSTEMS BUILDING NEWS, published monthly by W.R.C. Smith Publishing Co., 1760 Peachtree Rd., N.W., Atlanta, Georgia 30309. For the systems building fields in which modular or factory-manufactured components are mass-produced, transported to the building site, and assembled there. Covers finance, logistics, realty, critical path planning, etc. No cost to qualified readers.

Directories and Other Information Sources

ARCHITECTURAL ALUMINUM INDUSTRY DIRECTORY. Compilation of plants actively engaged in the manufacture of commercial, industrial, and residential architectural aluminum products. Over 800 listings by state, firm, and product line. Architectural Aluminum Manufacturers Association, 410 N. Michigan Ave., Chicago, Illinois 60611. Published annually, $25.

ARCHITECTURAL ALUMINUM MANUFACTURERS ASSOCIATION CERTIFIED PRODUCTS DIRECTORY. Quarterly listing of AAMA certified aluminum windows, sliding glass doors, and combination storm windows and doors and their manufacturers by specification designation. $2 per copy.

ARCHITECTURAL ALUMINUM MANUFACTURERS ASSOCIATION MEMBERSHIP DIRECTORY. Complete listing of AAMA national active and associate members, national officers, board of directors, member companies, voting representatives, and individual members. Also includes association staff, approved testing laboratories, committee rosters, and bylaws. Published annually; $100.

ASSOCIATED EQUIPMENT DISTRIBUTORS EDITION OF CONSTRUCTION EQUIPMENT BUYERS GUIDE. Listings of 1,700 U.S. and Canadian distributor companies. Contains names, addresses, and phone numbers of key people and lists the manufacturers each distributor represents. Includes product section with 600 standard product classifications by manufacturer. For AED members, $5; others, $10.

ASSOCIATED EQUIPMENT DISTRIBUTORS MEMBERSHIP DIRECTORY. Listings of AED U.S. and Canadian distributor and manufacturer members, trade press, and honorary members. Also included are officers, directors, staff, a local distributor association directory, plus AED constitution and bylaws and association committee members. AED members, $2.50; others, $5.

AUTOMATION IN HOUSING. List of major companies, their addresses, and product information for industrialized housing. Vance Publishing Group, 300 W. Adams St., Chicago, Illinois 60606. $1.50.

BUILDING CONSTRUCTION INFORMATION SOURCES. For researchers in the construction field and for contractors, builders, economists, architects, bankers, real estate operators, and developers. Tells how to get information on nearly 1,000 building topics. Gale Research Co., Book Tower, Detroit, Michigan 40226. $14.50.

DIRECTORY OF MODULAR HOUSING PRODUCERS. House & Home, 1221 Avenue of the Americas, New York, N.Y. 10020. $2.

MODULAR HOUSING. Listing of manufacturers of modular housing and an industry survey. Cahners Books, 89 Franklin St.,

Boston, Massachusetts 02110. $27.50.

HARDWARE AGE ANNUAL DIRECTORY ISSUE. Chilton Co., Chilton Way, Radnor, Pennsylvania 19089. No cost to qualified readers.

PRODUCERS COUNCIL GUIDE TO QUALITY CONSTRUCTION PRODUCTS. Directory of members of the Producers Council. Index of products by use and of council members of major products and equipment. Local representatives and supply sources are given.

SOUTHERN WHOLESALERS GUIDE. Compilation of the leading hardware and building supply manufacturers and the names, addresses, and telephone numbers of their Southern and Southwestern sales representatives. W.R.C. Smith Publishing Co., 1760 Peachtree Rd., N.W., Atlanta, Georgia 30309. No cost to qualified readers.

SWEET'S FILES: SWEET'S ARCHITECTURAL CATALOG FILE, SWEET'S CONSTRUCTION CATALOG SERVICES, SWEET'S INDUSTRIAL CONSTRUCTION CATALOG FILE, SWEET'S LIGHT CONSTRUCTION CATALOG FILE. Each compiled and issued annually by Sweet's Construction Division, McGraw-Hill Information Systems, 1221 Avenue of the Americas, New York, N.Y. 10020. No cost to qualified readers.

UNDERWRITERS LABORATORIES BUILDING MATERIALS DIRECTORY. Contains manufacturers who have demonstrated an ability to produce products, devices, or systems in accordance with the Laboratories' requirements. Includes a building materials list and a classified building materials index. Published yearly with quarterly supplements. Underwriters Laboratories Inc., 207 E. Ohio St., Chicago, Illinois 60611. $2.25.

UNDERWRITERS LABORATORIES FIRE RESISTANCE INDEX. Gives hourly ratings for beams, columns, floors, roofs, walls, and partitions, Published yearly with quarterly supplements. Underwriters Laboratories Inc., 207 E. Ohio St., Chicago, Illinois 60611. $2.75.

CONCRETE AND MASONRY—INCLUDING ASBESTOS-CEMENT, BRICK, BUILDING STONE, STRUCTURAL CLAY

Associations and Professional Groups

ALLIED STONE INDUSTRIES, Victor OOlitic Stone Co., Bloomington, Indiana 47402.

AMERICAN CONCRETE INSTITUTE, P.O. Box 19150, Redford Station, Detroit, Michigan 48219. Tel.: (313) 532-2600. Professional society of engineers, architects, and contractors involved with the design, construction, and maintenance of concrete structures. Publications:
Journal of the American Concrete Institute, monthly
Concrete Abstracts, bimonthly
Cement and Concrete Research, bimonthly

AMERICAN SOCIETY FOR CONCRETE CONSTRUCTION, P.O. Box 85, Addison, Illinois 60101.

ASBESTOS-CEMENT PRODUCTS ASSOCIATION, Room 2000, 521 Fifth Ave., New York, N.Y. 10017. Tel.: (212) 972-0500. Group of manufacturers of asbestos-cement flat and corrugated sheets and other building products.

BARRE GRANITE ASSOCIATION, 51 Church St., Barre, Vermont 05641. Tel.: (802) 476-4131. Group of manufacturers of granite for building stone and road aggregate. Publication:
Barre Life, quarterly

BRICK INSTITUTE OF AMERICA, 1750 Old Meadow Rd., McLean, Virginia 22101. Tel.: (703) 893-4010. Manufacturers of clay, brick and tile. Institute has a large library on masonry construction. Publication:
Brick & Tile, six times a year

BUILDING STONE INSTITUTE, 420 Lexington Ave., New York, N.Y. 10017. Tel.: (212) 532-9477. Composed of quarry owners and contractors. Publication:
Building Stone News, monthly

CELLULAR CONCRETE ASSOCIATION, 715 Boylston St.,

Boston, Massachusetts 02116. Tel.: (617) 266-6800. Organization of manufacturers and applicators of cellular concrete roof decks, floor fills, and structural work. Publications:

Newsletter
Membership Directory

CEMENT AND CONCRETE ASSOCIATION, Wexham Springs, Slough SL3 6PL, Bucks, England. Composed of seven cement companies in Great Britain. Offers technical information and advice.

CONCRETE PLANT MANUFACTURERS BUREAU, 900 Spring St., Silver Spring, Maryland 20910. Tel.: (301) 587-1400.

CONCRETE REINFORCING STEEL INSTITUTE, 180 N. LaSalle St., Room 2110, Chicago, Illinois 60601. Tel.: (312) 372-5059. Group of producers of steel products for concrete reinforcement. Provides information on design and construction techniques. Publication:

Shop Talk, quarterly

EXPANDED SHALE, CLAY, AND SLATE INSTITUTE, 1041 National Press Bldg., Washington, D.C. 20004. Tel.: (202) 783-1669. Technical organization supported by producers of expanded shale, clay, or slate by the rotary kiln process. Publication:

Expanded Shale Lightweight Concrete Facts Magazine, quarterly

EXPANSION JOINT MANUFACTURERS ASSOCIATION, 331 Madison Ave., New York, N.Y. 10017. Tel.: (212) 661-2050.

FACING TILE INSTITUTE, 111 E. Wacker Dr., Chicago, Illinois 60601. Tel.: (312) 644-6610. Group of manufacturers of glazed brick and tile.

FLEXICORE MANUFACTURERS ASSOCIATION, 2116 Jefferson St., Madison, Wisconsin 53711. Tel.: (608) 251-3077. Group of manufacturers of prestressed concrete products.

GYPSUM ASSOCIATION, 201 N. Wells St., Chicago, Illinois 60606. (See listing under BUILDINGS AND BUILDING

PRODUCTS in this section.)

INDIANA LIMESTONE INSTITUTE OF AMERICA, Stone City National Bank Bldg., Suite 400, Bedford, Indiana 47421. Tel.: (812) 275-4426. Producer group provides information on properties and uses of Indiana limestone.

INTERNATIONAL MASONRY INSTITUTE, 823 15th St., N.W., Washington, D.C. 20005. Tel.: (202) 783-3908. Organization of labor and contractor groups involved in masonry construction. Publication:

Masonry Promotion News, quarterly

LIGHTWEIGHT AGGREGATE PRODUCERS ASSOCIATION, 546 Hamilton St., Allentown, Pennsylvania 18101. Tel.: (215) 435-9687. Group fosters use of lightweight aggregates in concrete and masonry construction. Publication:

LAPA Newsletter

MARBLE INSTITUTE OF AMERICA, 1984 Chain Bridge Rd., McLean, Virginia 22101. Tel.: (703) 893-5800. Association of producers and contractors of building marble. Publishes design details for architects and others.

MASONRY INSTITUTE OF AMERICA, 2550 Beverly Blvd., Los Angeles, California 90057.

MO-SAI INSTITUTE INC., 110 Social Hall Ave., Salt Lake City, Utah 84111. Tel.: (801) 328-8663. Seventeen corporate members which produce precast concrete and stone products.

NATIONAL ASSOCIATION OF MARBLE PRODUCERS, 1984 Chain Bridge Rd., McLean, Virginia 22101.

NATIONAL BUILDING GRANITE QUARRIES ASSOCIATION, North State St., Concord, New Hampshire 03301. Group of manufacturers of building granite. Publication:

Granite in Architecture, annual

NATIONAL CONCRETE MASONRY ASSOCIATION, 1800 N. Kent St., Arlington, Virginia 22209. Tel.: (703) 524-0815. Composed of concrete block manufacturers and material

suppliers. Publication:
C/M Newsletter, monthly

NATIONAL CRUSHED STONE ASSOCIATION, 1415 Elliot Place, N.W., Washington, D.C. 20007. Tel.: (202) 333-1536. Group of manufacturers and processors of crushed stone for highway construction.

NATIONAL PRECAST CONCRETE ASSOCIATION, 2201 E. 46th St., Indianapolis, Indiana 46205. Tel.: (317) 253-0486. Manufacturers organization promotes the use of precast concrete by helping to develop high standards of quality and by exchanging information. Publication:
NPCA Newsletter, quarterly
Directory of Members and Precast Concrete Products

NATIONAL READY MIXED CONCRETE ASSOCIATION, 900 Spring St., Silver Spring, Maryland 20910. Tel.: (301) 587-1400. Organization of ready-mixed concrete producers.

NATIONAL SAND AND GRAVEL ASSOCIATION, 900 Spring St., Silver Spring, Maryland 20910. Tel.: (301) 587-1400. Organization of sand and gravel producers.

NATIONAL SLAG ASSOCIATION, 300 S. Washington St., Alexandria, Virginia 22314. Organization of companies involved in slag processing. Functions as a consulting service for field work and in preparing specifications. Conducts research in processing, quality control, and application of slag.

PORTLAND CEMENT ASSOCIATION, Old Orchard Rd., Skokie, Illinois 60076. Organization of cement producers involved with research and development in concrete technology, structural design, and construction techniques.

PRESTRESSED CONCRETE INSTITUTE, 20 N. Wacker Dr., Chicago, Illinois 60606. Tel.: (312) 346-4071. Membership of 1,600 manufacturers, engineers, and others who design and construct with prestressed concrete. Publications:
Journal of the Prestressed Concrete Institute, bimonthly
PC Items, monthly

REINFORCED CONCRETE RESEARCH COUNCIL, 3129 Civil Engineering Bldg., Urbana, Illinois 61801. Tel.: (312) 966-6200. Sponsors research in reinforced concrete. Publication:
RCRC Bulletin

WIRE REINFORCEMENT INSTITUTE, 7900 Westpark Dr., McLean, Virginia 22101. Members are manufacturers of steel wire fabric for concrete reinforcement. Publication:
Wire Journal, monthly

Periodicals

BRICK AND CLAY RECORD, published monthly by Cahners Publishing Co., 5 S. Wabash Ave., Chicago, Illinois 60603. For those engaged in mining and manufacturing of structural clay products: brick, refractories, clay pipe, and expanded clay and shale aggregates. No cost for qualified readers; $5 per year for others.

CONCRETE CONSTRUCTION, published monthly by Concrete Construction Publications Inc., P.O. Box 555, Elmhurst, Illinois 60126. For general contractors, concrete contractors, architects, and engineers responsible for the design, engineering, and construction of buildings, highways, and public works in connection with the specification, production, handling, forming, reinforcing, placing, finishing, and curing of concrete at the job site in its unhardened state. No cost to qualified readers; $5 per year for others.

CONCRETE INDUSTRIES YEARBOOK, published annually by Pit and Quarry Publications Inc., 105 W. Adams St., Chicago, Illinois 60607. For purchasing officials in the concrete industries. Covers plant design, plant operating data, new equipment and product application. No cost to qualified readers.

CONCRETE PRODUCTS, published monthly by Maclean-Hunter Publishing Corp., 300 W. Adams St., Chicago, Illinois 60606. For those in the concrete products field responsible for the manufacturing of concrete block, ready-mix concrete, concrete pipe, precast concrete and prestressed concrete. No cost to qualified readers; $10 per year for others.

MODERN CONCRETE, published monthly by Pit and Quarry Publications, 105 W. Adams St., Chicago, Illinois 60603. For producers of ready-mixed concrete, block, pipe, precast and prestressed concrete, other concrete units and contractors, engineers and architects allied with the production of concrete. No cost to qualified readers; $2 per year for others. Yearbook issued annually.

PIT & QUARRY, published monthly by Pit and Quarry Publications, 105 W. Adams St., Chicago, Illinois 60603. For those who specify or buy equipment, supplies, and services for the mining, quarrying, and processing of nonmetallic minerals. Includes producers of crushed stone, sand, gravel, cement, lime, gypsum, slag, lightweight aggregates, etc. No cost to qualified readers; $3 per year for others.

ROCK PRODUCTS, published monthly by MacLean-Hunter Publishing Co., 300 W. Adams St., Chicago, Illinois 60606. Supplies operating, technical, and business information to producers in the construction raw materials segment of the nonmetallic minerals industry. Includes mining, quarrying, and processing of portland cement, lime, gypsum, sand, gravel, crushed stone, slag, and the lightweight aggregates. No cost to qualified readers; $4 per year for others. Special issues: Cement, Lime, Aggregate Plant, and Forecast and Buyer's Guide.

Directories

AMERICAN CONCRETE INDUSTRY DIRECTORY. Listing of engineers, architects, manufacturers, and others who are involved in the concrete industry. (For members only.) American Concrete Institute, P.O. Box 4754, Redford Station, Detroit, Michigan 48219.

DIRECTORY OF BRICK AND TILE MANUFACTURERS IN THE AMERICAN STRUCTURAL CLAY PRODUCTS INDUSTRY. Listing of 362 manufacturers of brick and tile. $1. Structural Clay Products Institute, 1750 Old Meadow Rd., McLean, Virginia 22101.

DIRECTORY OF BRICK MANUFACTURERS. Convenient geographic and alphabetic listing of brick manufacturers in the

U.S. $2. Brick Institute of America, 1750 Old Meadow Rd., McLean, Virginia 22101.

PIT & QUARRY HANDBOOK AND PURCHASING GUIDE. Annual production and engineering reference manual for purchasers of cement, crushed stone, sand, gravel, slag, lime, gypsum, etc. No cost to qualified readers. Pit and Quarry Publications, 105 W. Adams St., Chicago, Illinois 60603.

ELECTRICAL—INCLUDING LIGHTING, SECURITY SYSTEMS, APPLIANCES, ENERGY, AND ENVIRONMENT

Associations and Professional Groups

AMERICAN HOME LIGHTING INSTITUTE, 230 N. Michigan Ave., Chicago, Illinois 60601. Tel.: (312) 236-7796. Group of manufacturers and distributors of lighting fixtures concerned with good residential lighting practice. Publication:
Light Rays, monthly

ASSOCIATION OF HOME APPLIANCE MANUFACTURERS, 20 N. Wacker Dr., Chicago, Illinois 60606. Tel.: (312) 236-2921.

ELECTRIC ENERGY ASSOCIATION, 90 Park Ave., New York, N.Y. 10016. Tel.: (212) 986-4154. Organization of more than 100 electric utilities and others. Promotes the interests of the utility industry. Publication:
EEA News, monthly

FLUORESCENT LIGHTING INSTITUTE, 101 Park Ave., New York, N.Y. 10017. Tel.: (212) 684-3160. Fosters the use of cold-cathode fluorescent lighting through education and research.

LIGHTNING PROTECTION INSTITUTE, 122 W. Washington Ave., Madison, Wisconsin 53703. Tel.: (808) 225-4223. Promotes lightning protection, standards of quality, and public education.

MANUFACTURERS OF ILLUMINATION PRODUCTS, Room 307, 158-11 Jewel Ave., Flushing, N.Y. 11365. Tel.: (212) 591-1100. Association of manufacturers of lamps and lighting fixtures.

NATIONAL AUTOMATIC SPRINKLER AND FIRE CONTROL ASSOCIATION, 2 Holland Ave., White Plains, N.Y. 10603. Tel.: (914) 428-2897. Association of manufacturers and contractors. Participates in fire protection and research and building code development. Publication:
News Bulletin, quarterly

NATIONAL BURGLAR AND FIRE ALARM ASSOCIATION, 1225 Connecticut Ave., N.W., Washington, D.C. 20036. Organization of manufacturers of alarm systems and contractors who install the equipment. Publications:
NBFAA Signal, quarterly
Signal Gram Newsletter, monthly

NATIONAL ELECTRICAL MANUFACTURERS ASSOCIATION, 155 E. 44th St., New York, N.Y. 10017. Tel.: (212) 682-1500. Organization of 500 manufacturers of electrical equipment. Has participated in developing electrical codes and many product standards. Publication:
NEMA Report, bimonthly

SAFE MANUFACTURERS NATIONAL ASSOCIATION, 366 Madison Ave., New York, N.Y. 10017. Tel.: (212) 682-2925.

SECURITY EQUIPMENT INDUSTRY ASSOCIATION, 360 N. Michigan Ave., Chicago, Illinois 60601. Tel.: (312) 332-2833. Group of manufacturers and distributors of security equipment.

Periodicals

EE—ELECTRICAL EQUIPMENT, published monthly by Sutton Publishing Co., 172 S. Broadway, White Plains, N.Y. 10605. Describes new electrical, electronic, and electromechanical equipment, components, accessories, materials, and related products and services. No cost to qualified readers.

ELECTRICAL APPARATUS, published monthly by Barks Publications, 360 N. Michigan Ave., Chicago, Illinois 60601. For those who sell, service, and remanufacture electrical equipment (motors, generators, transformers, and related controls, replacement parts, and supplies). No cost to qualified readers; $10 per year for others.

ELECTRICAL CONSTRUCTION AND MAINTENANCE, published monthly by McGraw-Hill Publications, 1221 Avenue of the Americas, New York, N.Y. 10020. For consulting electrical engineers, electrical contractors, and industrial and institutional electrical departments. $5 per year.

ELECTRICAL DISTRIBUTOR, published monthly by the National Association of Electrical Distributors, 111 Prospect St., Stamford, Connecticut 06901. No cost to qualified readers; $2 per year for others.

ELECTRICAL SOUTH, published monthly by Rickard Publishing Co., 1760 Peachtree Rd., N.W., Atlanta, Georgia 30309. Covers two fields: (1) design and installation of interior wiring systems and their associated equipment such as lighting, electrical heating; (2) design, installation, and testing of electric utility facilities such as transmission, distribution, metering, substations, but excluding generating facilities. No cost to qualified readers; $15 per year for others.

ELECTRICAL WHOLESALING, published monthly by McGraw-Hill Publications, 1221 Avenue of the Americas, New York, N.Y. 10020. For owners, corporate officials, managers, and sales personnel of electrical wholesale distributors. $4 per year.

ELECTRIC UTILITY PRODUCT NEWS, published quarterly by Cahners Publishing Co., 221 Columbus Ave., Boston, Massachusetts 02116. New product developments, applications, and information from electrical equipment suppliers and from the electric utility field. No cost to qualified readers.

ELECTRICITY IN BUILDING, published monthly by Electrical Information Publications, 2132 Fordem Ave., P.O. Box 1648, Madison, Wisconsin 53701. For home builders, electrical contractors, and others involved in construction or financing for the electric-home market. $3.50 per year.

INSULATION/CIRCUITS, published monthly by Lake Publishing Corp., Box 159, 700 Peterson Rd., Libertyville, Illinois 60048. For those involved in design, development, specifying, purchasing, assembly, and production/manufacturing work. No cost to qualified readers.

MIDWEST ELECTRICAL NEWS, published monthly by Rickard Publishing Co., Room 1713, 20 N. Wacker Dr., Chicago, Illinois 60606. Covers the activities of the people, companies, and associations that make up the electrical industry in the Midwest. No cost to qualified readers; $5 per year for others.

MODERN STORES AND OFFICES, published in even months by B.J. Martin Co., 20 N. Wacker Dr., Chicago, Illinois 60606. Covers general, exterior, and display lighting; comfort heating; electric signs; store fronts and entrances; air conditioning, electric wiring; snow melting; water heating. $6 per year.

NEW ENGLAND ELECTRICAL NEWS, published monthly by the Gleason Co., 1022 Morrissey Blvd., Boston, Massachusetts 02122. Feature articles describing electrical installations in New England with special emphasis on new materials and techniques. No cost to qualified readers; $5 per year for others.

Directories

CEE—CLASSIFIED DIRECTORY OF ELECTRICAL PRODUCTS, TOOLS, AND EQUIPMENT, published annually by Sutton Publishing Co., 172 S. Broadway, White Plains, N.Y. 10605. Products installed and used in the electrical construction market. No cost to qualified readers; $25 per copy for others.

EC&M'S ELECTRICAL PRODUCT YEARBOOK, published annually by McGraw-Hill Publications, 1221 Avenue of the Americas, New York, N.Y. 10020. Photos and detailed information on new products used in electrical systems in industrial, commercial, institutional, and residential buildings. No cost to qualified readers.

INTERNATIONAL DIRECTORY OF ELECTRICITY SUPPLIERS, published annually by McGraw-Hill Publications, 1221 Avenue of the Americas, New York, N.Y. 10020. Contains information on personnel and facilities of over 600 electricity suppliers, including data on installed capacity, energy production and sales, number of customers and employees. For each country: a short resume of the electrical industry, name of capital city, area, population, language, currency with U.S. dollar equivalent, number of telephones and television sets, electric current characteristics of principal cities. No cost to qualified readers; $75 per copy for others.

LIGHT AGE DIRECTORY, published annually by Rosenthal and Smythe, Inc., 1115 Clifton Ave., Clifton, N.J. 07103. Lists trade sources for portable lamps, lamp shades, bulbs, and lighting fixtures. No cost to qualified readers; $2 per copy for others.

LIGHTING EQUIPMENT BUYERS GUIDE, published by W.R.C. Smith, 1760 Peachtree Rd., N.W., Atlanta, Georgia 30309. $5.

MIDWEST ELECTRICAL BUYERS GUIDE, published annually by Rickard Publishing Co., 20 N. Wacker Dr., Chicago, Illinois 60606. Contains over 1,000 electrical product classifications, with manufacturers of each product. Also lists Midwestern sales office and sales agents for each manufacturer. No cost to qualified readers; $5 for others.

NEW ENGLAND ELECTRICAL BLUE BOOK, published annually by the Gleason Co., 1022 Morrissey Blvd., Boston, Massachusetts 02122. Lists electrical manufacturers that maintain sales representation in New England, with their products. No cost to qualified readers; $15 per copy for others.

UNDERWRITERS LABORATORIES ELECTRICAL CONSTRUCTION MATERIALS LIST, published yearly, with quarterly supplements, by Underwriters Laboratories, Inc., 207 E. Ohio St., Chicago, Illinois 60611. For rated equipment used in ordinary locations. $3.

UNDERWRITERS LABORATORIES HAZARDOUS LOCATION EQUIPMENT LIST, published yearly, with quarterly supplements, by Underwriters Laboratories, Inc., 207 E. Ohio St., Chicago, Illinois 60611. $1.

FINISHES—INCLUDING PAINT, WALLPAPER, PLASTER, FLOOR COVERING, TILE

Associations and Professional Groups

ASPHALT AND VINYL ASBESTOS TILE INSTITUTE, 101 Park Ave., New York, N.Y. 10017. Tel.: (212) 686-3937. Manufacturers of asphalt and vinyl asbestos tile are members of the institute.

CERAMIC TILE INSTITUTE, 700 W. Virgil St., Los Angeles, California 90029. Tel.: (213) 660-1911.

INTERNATIONAL COUNCIL FOR LATHING AND PLASTERING, 221 N. LaSalle St., Chicago, Illinois 60601. Tel.: (312) 346-1862. Council is made up of lathing and plastering contractors, labor unions, and manufacturers.

METAL LATH ASSOCIATION, 221 N. LaSalle St., Chicago, Illinois 60601. Tel.: (312) 346-8717. Association of manufacturers of metal lath. Conducts research in fire-resistant and sound transmission properties of metal lath and plaster construction. Publication:
Metal Lath News, quarterly

NATIONAL ASSOCIATION OF FLOOR COVERING WHOLESALERS, 221 N. LaSalle St., Chicago, Illinois 60601. Tel.: (312) 332-7127.

NATIONAL DECORATING PRODUCTS ASSOCIATION, Dielman Industrial Dr., St. Louis, Missouri 63132. Tel.: (314) 962-9860. Group of retailers and distributors of paint and wallpaper. Publications:
Decorating Retailer, monthly
Directory

NATIONAL PAINT AND COATINGS ASSOCIATION, 1500 Rhode Island Ave., N.W., Washington, D.C. 20005. Tel.: (202) 462-6272. Manufacturers of paints and coatings. Committees include Fire-Retarding Paints, High Performance Architectural Coatings, Roof Coatings, and Sealants. Publication:
Coatings

NATIONAL TERRAZZO AND MOSAIC ASSOCIATION, 716 Church St., Alexandria, Virginia 22314. Tel.: (703) 836-6765. Provides information on terrazzo and mosaic work and development of conductive flooring. Publications:

Terrazzo Topics, monthly
Terrazzo Trends, semiannual
Directory, annual

PORCELAIN ENAMEL INSTITUTE, 1900 L St., N.W., Washington, D.C. 20036. Tel.: (202) 296-0450. Organization of manufacturers of porcelain enamel products for appliances, architectural purposes, signs, etc. Publication:

PEI Newsletter, monthly

RESILIENT FLOORING AND CARPET ASSOCIATION, Box 11082, Oakland, California 94611. Tel.: (415) 526-8668.

RESILIENT TILE INSTITUTE, 26 Washington St., East Orange, N.J. 07017.

STAINED AND LEADED GLASS ASSOCIATION, 40 W. 13th St., New York, N.Y. 10011. Tel.: (212) 675-0400.

STAINED GLASS ASSOCIATION OF AMERICA, 1125 Wilmington Ave., St. Louis, Missouri 63111. Tel.: (314) 353-5683. Group of stained glass manufacturers and craft studios. Publication:

Stained Glass Journal, quarterly

STEEL STRUCTURES PAINTING COUNCIL, 4400 Fifth Ave., Pittsburgh, Pennsylvania 15213. Tel.: (412) 621-1100. Group of organizations interested in reducing corrosion in steel structures and improving maintenance techniques. Publication:

Steel Structures Painting Bulletin

STUCCO MANUFACTURERS ASSOCIATION, 14006 Ventura Blvd., Suite 204, Sherman Oaks, California 91403. Tel.: (213) 789-8733. Twenty-five manufacturers of stucco products and accessories.

TILE COUNCIL OF AMERICA, Box 326, Princeton, N.J.

08540. Tel.: (609) 921-7050. Composed of 25 companies that manufacture ceramic tile.

WALLCOVERINGS COUNCIL, INC., 969 Third Ave., New York, N.Y. 10022. Tel.: (212) 759-0950.

WALLPAPER INSTITUTE, 969 Third Ave., New York, N.Y. 10022. Tel.: (212) 759-0950.

WOOD AND SYNTHETIC FLOORING INSTITUTE, 1201 Waukegan Rd., Glenview, Illinois 60025. Tel.: (312) 724-7700. Group of installation contractors of wood and synthetic flooring and industry suppliers and manufacturers.

Periodicals

CONTRACT FLOOR COVERINGS, published quarterly by Charleson Publishing Co., 630 Third Ave., New York, N.Y. 10017. For contract specifiers, including architects, interior designers, contract furnishers, major home builders, and government agencies that purchase or specify carpets, rugs, resilient flooring, and allied floor covering products and services. No cost to qualified readers; $3 per copy for others.

FLOOR COVERING WEEKLY, published by Bart Publications, Empire State Bldg., New York, N.Y. 10001. For retailers, furniture and department stores, decorators, designers, lumber yards, contractors, wholesalers, and manufacturers. $6 per year.

FLOORING, published monthly by Harcourt Brace Jovanovich, 757 Third Ave., New York, N.Y. 10017. For dealer-contractors and distributors engaged in the sale and installation of floor, wall, and related interior surfacing products, including resilient flooring, carpets, rugs, ceramic tile, hardwood flooring, plastic laminates, ceiling systems, seamless flooring, terrazzo and industrial flooring. No cost to qualified readers; $5 per year for others.

INSTALLATION SPECIALIST, published six times a year by Specialist Publications, Suite 311, 15130 Ventura Blvd., Sherman Oaks, California 91403. For wall covering and counter top installers, workrooms and retail contract shops, and contractors. No cost to qualified readers; $4 per year for others.

MODERN FLOOR COVERINGS, published monthly by Charleson Publishing Co., 630 Third Ave., New York, N.Y. 10017. For the retailer, contractor, wholesale distributor, and manufacturer of carpets and rugs, resilient floor products (vinyl, linoleum, asphalt tile, etc.), and allied products and accessories. No cost to qualified readers.

TILE AND DECORATIVE SURFACES, published seven times a year by Tile and Architectural Ceramics Publications, 3421 Ocean View Rd., Glendale, California 91208. For the building construction industry, including tile contractors, terrazzo contractors, general contractors, architects, specification writers, and interior decorators. No cost to qualified readers; $4 per year for others. December Catalog and Directory Issue.

WESTERN PAINT REVIEW, published monthly by Linley Publishing Co., 1333 W. 8th St., Los Angeles, California 90057. For paint dealers and painting contractors. No cost to qualified readers; $4 per year for others.

Directory

DECORATING RETAILER'S DIRECTORY OF THE WALLCOVERINGS INDUSTRY, published annually by the National Decorating Products Association, 9334 Dielman Industrial Dr., St. Louis, Missouri 63132. Directory of the nearest stocking sources, wholesale or direct, of over 1,000 wallcovering sample books. Codifies pertinent data on manufacturers, their wallcovering lines, collection characteristics, book serial numbers, and expiration dates.

LUMBER—INCLUDING WOOD CONSTRUCTION AND PRESERVATION

Associations and Professional Groups

AMERICAN INSTITUTE OF TIMBER CONSTRUCTION, 333 W. Hampden Ave., Englewood, Colorado 80110. Tel.: (303) 761-3212. Group of manufacturers of glued laminated timber and engineers. Developed standards for timber design and construction. Large library.

AMERICAN PLYWOOD ASSOCIATION, 1119 A St., Tacoma, Washington 98401. Tel.: (206) 272-2283. Trade organization representing most of the nation's softwood plywood manufacturers. Conducts research to improve plywood performance, inspects and tests plywood for high quality, and promotes new information. Publication:

Plywood Statistics, weekly and monthly

AMERICAN WOOD COUNCIL, 1619 Massachusetts Ave., N.W., Washington, D.C. 20036. Tel.: (202) 265-7766. Association formed to promote acceptance of wood products for the residential market.

AMERICAN WOOD PRESERVERS ASSOCIATION, 1625 Eye St., N.W., Washington, D.C. 20006. Tel.: (202) 347-3282. Publication:

AWPA Book of Standards, annual

AMERICAN WOOD PRESERVERS INSTITUTE, 1651 Old Meadow Rd., McLean, Virginia 22101. Tel.: (703) 893-4005. Publication:

Wood Preserving, monthly

APPALACHIAN HARDWOOD MANUFACTURERS, INC., Room 408, NCNB Bldg., 164 S. Main St., High Point, North Carolina 27260. Organization of almost 100 manufacturers of Appalachian hardwoods. Publication:

Appalachian Hardwood Year Book

AROMATIC RED CEDAR CLOSET LINING MANUFACTURERS ASSOCIATION, 221 N. LaSalle St., Chicago, Illinois 60601. Tel.: (312) 372-7090. Organization of four manufacturers.

CALIFORNIA REDWOOD ASSOCIATION, 617 Montgomery St., San Francisco, California 94111. Organization of the principal redwood producers. Concerned with protection of future growth, utilization of the present crop, research, marketing, and promotion. Publication:

Redwood News, quarterly

FINE HARDWOODS—AMERICAN WALNUT ASSOCIATION, 666 N. Lake Shore Dr., Chicago, Illinois 60611. Tel.: (312)

944-6162. Association of manufacturers of hardwood veneer and lumber. Publication:
Newsletter, monthly

FOREST PRODUCTS RESEARCH SOCIETY, 2801 Marshall Court, Madison, Wisconsin 53705. Tel.: (608) 231-1361. Publications:
Forest Products Journal, monthly
Wood Science, quarterly

HARDWOOD DIMENSION MANUFACTURERS ASSOCIATION, 3813 Hillsboro Rd., Nashville, Tennessee 37215. Tel.: (615) 269-3254. Group of more than 40 manufacturers of hardwood furniture and cabinet components, moldings, and stair treads and risers.

HARDWOOD PLYWOOD MANUFACTURERS ASSOCIATION, P.O. Box 6246, Arlington, Virginia 22206. Tel.: (703) 671-6262. Composed of manufacturers of stock, cut-to-size, and prefinished hardwood plywood, and manufacturers of laminated hardwood block flooring. Maintains a testing laboratory for product research, quality control, and inspection services.

MAPLE FLOORING MANUFACTURERS ASSOCIATION, 424 Washington Ave., Oshkosh, Wisconsin 54901. Tel.: (414) 233-1920. Group of manufacturers of maple, birch, and beech flooring.

NATIONAL BUILDING MATERIAL DISTRIBUTORS ASSOCIATION, 221 N. LaSalle St., Chicago, Illinois 60601. Tel.: (312) 332-7127.

NATIONAL BUILDING PRODUCTS ASSOCIATION, 120-44 Queens Blvd., Kew Gardens, N.Y. 11415. Tel.: (212) 261-0510.

NATIONAL FOREST PRODUCTS ASSOCIATION, 1619 Massachusetts Ave., N.W., Washington, D.C. 20036. Tel.: (202) 332-1050. Association of forest products manufacturers representing major segments of the building materials industry. Publication:
National Forest Industries Newsletter, weekly

NATIONAL HARDWOOD LUMBER ASSOCIATION, 332 S. Michigan Ave., Chicago, Illinois 60604. Tel.: (312) 427-2811. Organization of lumber and veneer producers and distributors. Publications:

Newsletter, monthly
Yearbook

NATIONAL LUMBER AND BUILDING MATERIAL DEALERS ASSOCIATION, 1990 M St., N.W., Washington, D.C. 20036. Tel.: (202) 872-8860. Publication:

National News, monthly

NATIONAL OAK FLOORING MANUFACTURERS ASSOCIATION, 814 Sterick Bldg., Memphis, Tennessee 38103. Tel.: (901) 526-5016.

NATIONAL WOODWORK MANUFACTURERS ASSOCIATION, 400 W. Madison Ave., Chicago, Illinois 60606. Tel.: (312) 782-6232.

NORTH AMERICAN WHOLESALE LUMBER ASSOCIATION, 180 Madison Ave., New York, N.Y. 10016. Tel.: (212) 532-9161. Organization of wholesale distributors of lumber.

NORTHEASTERN LUMBER MANUFACTURERS ASSOCIATION, 13 South St., Glens Falls, N.Y. 12801. Tel.: (518) 793-3411. Organization of manufacturers of hardwood and softwood lumber.

NORTHERN HARDWOOD AND PINE MANUFACTURERS ASSOCIATION, 501 Norther Bldg., Green Bay, Wisconsin 54301. Tel.: (414) 432-9161. Organization of manufacturers of hardwood and softwood lumber. Publications:

Newsletter, quarterly
Buyers Guide, annual

NORTHWESTERN LUMBERMEN INC., 5309 Vernon Ave., Minneapolis, Minnesota 55436. Tel.: (612) 929-8553. Organization of retail lumber dealers in Iowa, the Dakotas, and Minnesota.

PHILIPPINE MAHOGANY ASSOCIATION, Box 3362, Tacoma, Washington 98499. Tel.: (206) 588-9050. Organization of importers. Provides information on properties, finishes, and uses of Philippine Mahogany.

PLASTICS IN CONSTRUCTION COUNCIL, 250 Park Ave., New York, N.Y. 10017. Tel.: (212) 687-2675.

PONDEROSA PINE WOODWORK, 1500 Yeon Bldg., Portland, Oregon 97204. Tel.: (503) 224-3930. Organization of manufacturers of windows, doors and millwork.

REDWOOD INSPECTION SERVICE, 617 Montgomery St., San Francisco, California 94111.

RED CEDAR SHINGLE AND HANDSPLIT SHAKE BUREAU, 5510 White Bldg., Seattle, Washington 98101. Tel.: (216) 623-4881. Organization of manufacturers.

SOUTHERN CYPRESS MANUFACTURERS ASSOCIATION, Box 5816. Jacksonville, Florida 32207. Tel.: (904) 398-4224. Group of manufacturers of red cypress lumber.

SOUTHERN FOREST INSTITUTE, One Corporate Square, N.E., Atlanta, Georgia 30329. Tel.: (404) 633-5137.

SOUTHERN FOREST PRODUCTS ASSOCIATION, Box 52468, New Orleans, Louisiana 70152. Tel.: (504) 834-8544. Organization of Southern pine lumber manufacturers.

SOUTHERN HARDWOOD LUMBER MANUFACTURERS ASSOCIATION, 805 Sterick Bldg., Memphis, Tennessee 38103. Tel.: (901) 525-8221.

STRUCTURAL WOOD FIBER PRODUCTS ASSOCIATION, 5028 Wisconsin Ave., N.W., Washington, D.C. 20016. Tel.: (202) 686-2895. Organization of manufacturers of roof decks, form boards, and acoustical material from wood fiber.

TRUSS PLATE INSTITUTE, 919 18th St., N.W., Washington, D.C. 20006. Tel.: (202) 293-2599. Association of manufacturers of connectors for structural joints in wood and timber construction.

WESTERN RED AND NORTHERN WHITE CEDAR AS-

SOCIATION, P.O. Box 2786, New Brighton, Minnesota 55112.

WESTERN RED CEDAR LUMBER ASSOCIATION, Yeon Bldg., Portland, Oregon 97204. Tel.: (503) 224-3930. Association of Western red cedar manufacturers. Publication:
Concepts in Cedar, quarterly

WESTERN WOOD PRODUCTS ASSOCIATION, 1500 Yeon Bldg., Portland, Oregon 97204. Tel.: (503) 224-3930. Organization of lumber manufacturers of 12 Western states. Publication:
Plumb Line, weekly

Periodicals

AMERICAN BUILDING SUPPLIES, published monthly by W.R.C. Smith Publishing Co., 1760 Peachtree Rd., N.W., Atlanta, Georgia 30309. For lumber and building supply dealers and wholesalers.

CROW'S FOREST PRODUCTS DIGEST, published monthly by C.C. Crow Publications, Terminal Sales Bldg., Portland, Oregon 97205. Marketing news for marketers, distributors, and end users of wood products and related building supplies, including lumber and building materials dealers, prefab, mobile, or modular home manufacturers, lumber, millwork, and shingle manufacturers, hardboard, particleboard, and plywood manufacturers, lumber and building materials.wholesalers, plywood jobbers, and sash-and-door jobbers. No cost to qualified readers; $6 per year for others. Special issues: Plywood Guide (March); Buyers and Sellers Guide (every two years).

EASTERN BUILDING MATERIALS AND LUMBER TRADE JOURNAL, published bimonthly by Curtis Guild and Co., 88 Broad St., Boston, Massachusetts 02110. For retailers, wholesalers, and distributors of lumber and building materials in the New England and Middle Atlantic states. $4 per year.

ILLINOIS BUILDING NEWS, published monthly by the Illinois Lumber and Material Dealers Association, 620 Reisch Bldg., Springfield, Illinois 62701. For management and sales personnel in the retail lumber and building materials industry. $3.50 per year.

LUMBER CO-OPERATOR, published monthly by the Lumber Co-operator, Inc., 339 East Ave., Rochester, N.Y. 14604. Official publication of the Northeastern Retail Lumbermen's Association. For retail lumber and building material dealer-members in the Northeast. $5.50 per year.

MISSISSIPPI VALLEY LUMBERMAN, published monthly by Lumberman Publishing Co., 1100 Upper Midwest Bldg., Minneapolis, Minnesota 55401. For those in the retail and wholesale building materials industries in the North Central U.S.

NATIONAL SASH AND DOOR JOBBERS ASSOCIATION DIGEST, published monthly by Associations Publishing Inc., P.O. Box 17626, 726 Mt. Moriah Rd., Suite 106, Memphis, Tennessee 38117. For owners and management of manufacturing, jobbing (distribution), and retailing firms in the building products industry. No cost to qualified readers; $4 per year for others.

PLYWOOD & PANEL MAGAZINE, published monthly by the Saturday Evening Post Co., 1100 Waterway Blvd., Indianapolis, Indiana 46202. For persons concerned with production, sales, marketing, distribution, fabrication, and utilization of veneer, plywood, and other panel products. No cost to qualified readers; $6 per year for others.

RETAIL LUMBERMAN, official publication of Southwestern Lumbermen's Association and Mountain States Lumber Dealers Association. Published monthly by Retail Lumberman Publishing Co., 1102 Grand Ave., Kansas City, Missouri 64106. For retail lumber and building materials dealers. Covers industry innovations and improved business operating methods. $2 per year.

WESTERN LUMBER & BUILDING MATERIALS MERCHANT, published monthly by California Lumber Merchant Inc., 573 S. Lake Ave., Pasadena, California 91101. For retail lumber and building materials dealers in the 11 Western states. Also covers wholesaler and manufacturer news and operations. $5 per year.

Directories

DUN & BRADSTREET REFERENCE BOOK OF LUMBER

AND WOOD PRODUCTS INDUSTRIES. Directory of firms in the lumber business by size and sales ratings. Published by Dun and Bradstreet Inc., 99 Church St., New York, N.Y. 10007.

WHERE TO BUY HARDWOOD PLYWOOD AND VENEER. Hardwood Plywood Manufacturers Association, P.O. Box 6246, Arlington, Virginia 22206.

MECHANICAL—INCLUDING HVAC, PIPING, DRAINAGE, PLUMBING

Associations and Professional Groups

AIR-CONDITIONING AND REFRIGERATION INSTITUTE, 1815 N. Ft. Myer Dr., Arlington, Virginia 22209. Tel.: (703) 524-8800. National trade association of manufacturers of commercial and industrial air-conditioning and refrigeration equipment, machinery, parts, accessories, and allied products. Publication:
Kold-Fax Newsletter, quarterly

AIR DIFFUSION COUNCIL, 435 N. Michigan Ave., Chicago, Illinois 60611. Tel.: (312) 527-5494. Group of manufacturers of grilles, diffusers, and air control hardware.

AIR MOVING AND CONDITIONING ASSOCIATION, 30 W. University Dr., Arlington Heights, Illinois 60004. Tel.: (312) 394-0150. Group of manufacturers of air-conditioning and moving equipment. Conducts research and develops standards of performance.

AMERICAN BOILER MANUFACTURERS ASSOCIATION, 1500 Wilson Blvd., Arlington, Virginia 22209. Tel.: (703) 522-7298.

AMERICAN CONCRETE PIPE ASSOCIATION, 1501 Wilson Blvd., Arlington, Virginia 22209. Tel.: (703) 524-3939. Organization of manufacturers of concrete pipe. Provides information on properties and installation procedures. Publication:
Concrete Pipe News, bimonthly

AMERICAN CONCRETE PRESSURE PIPE ASSOCIATION, 1501 Wilson Blvd., Arlington, Virginia 22209. Tel.: (703) 524-

3939. Provides information on properties and standards for installation of concrete pressure pipe.

AMERICAN GAS ASSOCIATION, 1515 Wilson Blvd., Arlington, Virginia 22209. Tel.: (703) 524-2000. Publication:
AGA Monthly

AMERICAN PIPE FITTINGS ASSOCIATION, 26 Sixth St., Stamford, Connecticut 06905. Tel.: (203) 348-6459. Group of manufacturers of cast and malleable iron, brass, and steel pipe fittings.

BITUMINOUS PIPE INSTITUTE, 8 S. Michigan Ave., Chicago, Illinois 60603. Tel.: (312) 726-2030. Manufacturers of bituminous nonpressure pipe.

CAST IRON PIPE RESEARCH ASSOCIATION, 1211 W. 22nd St., Oak Brook, Illinois 60521. Tel.: (312) 654-2945. Group of manufacturers of cast iron pipe for water, sewer, and gas use. Publication:
Cast Iron Pipe News, quarterly

CAST IRON SOIL PIPE INSTITUTE, 2029 K St., N.W., Washington, D.C. 20006. Tel.: (202) 223-4536. Organization of manufacturers of soil pipe and fittings. Conducts research, develops standards, and distributes technical information.

COMPRESSED GAS ASSOCIATION, 500 Fifth Ave., New York, N.Y. 10036. Tel.: (212) 524-4796. Association of manufacturers and distributors of compressed and liquefied gases. Has developed safety standards for their use.

CONCRETE PIPE ASSOCIATION, 1501 Wilson Blvd., Arlington, Virginia 22209. Tel.: (703) 524-3939. Coordinates matters of interest to the American Concrete Pipe and American Concrete Pressure Pipe Associations. Publications:
Concrete Pipe News, bimonthly
Membership Directory, annual

CONTRACTORS PUMP BUREAU, 13975 Connecticut Ave., Silver Spring, Maryland 20906. Tel.: (301) 871-8988. Group of pump manufacturers sponsored by the Associated General Contractors of America, engaged in standardizing contractors' pumps.

COOLING TOWER INSTITUTE, 3003 Yale St., Houston, Texas 77018. Tel.: (713) 861-5328. Organization conducts research aimed at improving cooling tower performance; develops standards for construction and maintenance. Publication:
CTI News, quarterly

HOME VENTILATING INSTITUTE, 230 N. Michigan Ave., Chicago, Illinois 60601. Tel.: (312) 236-5822.

HYDRONIC INSTITUTE, 35 Russo Place, Berkeley Heights, N.J. 07922. Tel.: (201) 464-8200. Organization of manufacturers and contractors of hot water and steam heating and cooling devices.

INTERNATIONAL DISTRICT HEATING ASSOCIATION, 5940 Baum Square, Pittsburgh, Pennsylvania 15206. Organization of manufacturers and contractors involved in central-station space heating (steam and hot water) and air-conditioning (steam and chilled water). Publication:
District Heating, quarterly

MANUFACTURERS STANDARDIZATION SOCIETY OF THE VALVE AND FITTINGS INDUSTRY, 1815 N. Ft. Myer Dr., Arlington, Virginia 22209. Tel.: (703) 525-8526.

NATIONAL CLAY PIPE INSTITUTE, 350 W. Terra Cotta Ave., Crystal Lake, Illinois 60014. Tel.: (815) 459-3330. Manufacturers of clay sewer pipe, chimney pipe, and other clay products.

NATIONAL CORRUGATED STEEL PIPE ASSOCIATION, Chicago-O'Hare Aerospace Center, 4825 N. Scott St., Schiller Park, Illinois 60176. Tel.: (312) 678-5830. Group of manufacturers of corrugated metal pipe for drainage. Publishes manuals on installation of corrugated pipe. Publication:
NCSPA Newsletter, monthly

NORTH AMERICAN HEATING AND AIR-CONDITIONING WHOLESALERS ASSOCIATION, 1200 W. Fifth Ave., Co-

lumbus, Ohio 43212. Tel.: (614) 488-9769. Organization of wholesalers of heating and air-conditioning equipment. Publication:
Management Helps, monthly

NATIONAL WATER WELL ASSOCIATION, 88 E. Broad St., Columbus, Ohio 43215. Tel.: (614) 224-6241. Organization of the groundwater industry. Publication:
Water Well Journal, monthly

NATIONAL LP-GAS ASSOCIATION, 79 W. Monroe St., Chicago, Illinois 60603. Tel.: (312) 372-5484. Information arm of the LP-gas industry has experts available to supply information to architects and builders. Publication:
Newsletter, semimonthly

PIPE FABRICATION INSTITUTE, 1326 Freeport Rd., Pittsburgh, Pennsylvania 15238. Tel.: (412) 782-1624. National trade association whose members produce piping systems for electric power generating plants and other industries.

PLASTICS PIPE INSTITUTE, 250 Park Ave., New York, N.Y. Tel.: (212) 687-2675. Made up of manufacturers of plastic pipe.

PLUMBING AND DRAINAGE INSTITUTE, 1018 N. Austin Rd., Oak Park, Illinois 60302. Tel.: (312) 848-5797. Organization of manufacturers of plumbing fittings.

PLUMBING AND HEATING WHOLESALERS EMPLOYERS ASSOCIATION, 342 Madison Ave., New York, N.Y. 10017. Tel.: (212) 682-2930.

PLUMBING BRASS INSTITUTE, 221 N. LaSalle St., Chicago, Illinois 60601. Tel.: (312) 346-1862. Group of manufacturers of brass plumbing fittings.

PLUMBING-HEATING-COOLING INFORMATION BUREAU, 35 E. Wacker Dr., Chicago, Illinois 60601. Tel.: (312) 372-7331. Source of technical and business information; operated by manufacturers, contractors, and others.

VALVE MANUFACTURERS ASSOCIATION, 60 E. 42 St.,

New York, N.Y. 10017. Tel.: (212) 682-3640. Publication:
Valviews, quarterly

WATER AND WASTEWATER EQUIPMENT MANUFACTURERS ASSOCIATION, 744 Broad St., Newark, N.J. 07102. Tel.: (201) 622-0166. Publication:
News from WWEMA

Periodicals

AIR CONDITIONING, HEATING, AND REFRIGERATION NEWS, published weekly by Business News Publishing Co., P.O. Box 6000, 700 E. Maple, Birmingham, Michigan 48012. Covers manufacturing, sales, distribution, technology, and service. $10 per year.

CONTEMPORARY DESIGN (HEATING-PLUMBING-COOLING), published quarterly by Gas Magazines Inc., 1202 S. Park St., Madison, Wisconsin 53715. $2 per year.

DISTRICT HEATING, published quarterly by the International District Heating Association, 5940 Baum Square, Pittsburgh, Pennsylvania 15206. For producers and users of district heating services, and for those engaged in manufacturing and selling related equipment and supplies. $5 per year.

ELECTRIC COMFORT CONDITIONING JOURNAL, published monthly by Electrical Information Publications, 2132 Fordem Ave., P.O. Box 1648, Madison, Wisconsin 53701. For electric heating contractors, electrical contractors, heating and air conditioning contractors, consulting engineers, architects, builders, and electric utility comfort conditioning specialists. Reports on ideas for selling, designing, and installing electric heating and comfort conditioning systems, and on new developments in the industry. No cost to qualified readers; $3.75 per year for others. Special issue: Fact Book.

HEATING AND PLUMBING MERCHANDISER, published bimonthly by Master Plumber Publishing Co., P.O. Box 343, East Paterson, N.J. 07407. For contractors, wholesalers, specifying engineers, builders, mechanical contractors, and general contractors in the field of plumbing, heating, and cooling. Contains mostly product news. No cost to qualified readers.

REEVES JOURNAL, PLUMBING-HEATING-COOLING, published monthly by Miramar Publishing Co., 2048 Cotner Ave., Los Angeles, California 90025. No cost to qualified readers; $5 per year for others. Special issues: New Products and Equipment; Kitchen and Bath Remodeling.

REFRIGERATION, published monthly by John W. Yopp Publications, P.O. Box 7368, 770 Spring St., N.W., Atlanta, Georgia 30309. For owners and managers of ice, cold storage, and other refrigeration plants. No cost to qualified readers; $4 per year for others.

WHOLESALER, published monthly by Scott Periodicals Corp., 522 N. State Rd., Briarcliff Manor, N.Y. 10510. For wholesale distributors of plumbing, air conditioning, and refrigeration supplies.

Directories

AIR CONDITIONING, HEATING, AND REFRIGERATION NEWS DIRECTORY. Lists manufacturers and trade organizations connected with air conditioning, heating, and refrigeration equipment. Business News Publishing Co., P.O. Box 6000, Birmingham, Michigan 48012. $3.

AMERICAN ARTISAN DIRECTORY SECTION. List of manufacturers of HVAC, sheet metal, and dust control equipment. Keeney Division, Reinhold Publishing Corp., 10 S. LaSalle St., Chicago, Illinois 60603. $1.

CERTIFIED HOME VENTILATING PRODUCTS DIRECTORY. Listing of range hoods and wall and ceiling exhaust fans which were tested and rated for air delivery and sound emission by the Home Ventilating Institute, 230 N. Michigan Ave., Chicago, Illinois 60601.

DIRECTORY OF PRODUCTS LICENSED TO BEAR THE AMCA CERTIFIED RATINGS SEAL. Manufacturers are listed alphabetically under each product classification. Includes descriptions of all newly certified products and of any product that has lost its certification. Air Moving and Conditioning Association, 30 W. University Dr., Arlington Heights, Illinois 60004. Revised semiannually. No charge.

MECHANICAL PRODUCTS CATALOG, published annually by Hutton Publishing Co., 450 Plandome Rd., Manhasset, N.Y. 11030. For specifying mechanical engineers and mechanical contractors serving the fields of heating, air conditioning, ventilating, plumbing, and piping for commercial, institutional, industrial, and government buildings. In addition to manufacturers' catalogs, contains a product directory of manufacturers, trade name index, manufacturers' address list, and list of manufacturers' sales offices and representatives. No cost to qualified readers; $20 per year for others.

WHOLESALER PRODUCT DIRECTORY, published annually by Scott Periodicals Corp., 522 N. State Rd., Briarcliff Manor, N.Y. 10510. Comprehensive guide to who manufactures what in the plumbing, heating, cooling, piping, air conditioning, and refrigeration industries. No cost to qualified readers.

OTHER MATERIALS—INCLUDING ASPHALT, METALS, THERMAL AND ACOUSTIC INSULATION

Associations and Professional Groups

ACOUSTICAL AND INSULATING MATERIALS ASSOCIATION, 205 W. Touhy Ave., Park Ridge, Illinois 60068. Tel.: (312) 692-5178. Fourteen manufacturers, producing over 90 percent of the acoustical ceiling tile and insulation board in the U.S., constitute the membership. Association provides information to architects, engineers, contractors, builders, etc., on the use and properties of the products.

ALUMINUM ASSOCIATION, 750 Third Ave., New York, N.Y. 10017. Tel.: (212) 972-1800. Members represent the producers of aluminum in the U.S. Publications:
Aluminum Statistical Review, annual
About Aluminum, quarterly

AMERICAN HOT DIP GALVANIZERS ASSOCIATION, 1000 Vermont Ave., N.W., Washington, D.C. 20005. Tel.: (202) 628-4634.

AMERICAN INSTITUTE OF STEEL CONSTRUCTION, 101 Park Ave., New York, N.Y. 10017. Tel.: (212) 685-7374. Organization of fabricators and erectors of steel structures. Supports studies on structural steel design, erection, and maintenance. Publications:

AISC Engineering Journal, quarterly
AISC News, monthly
Modern Steel Construction, quarterly

AMERICAN IRON AND STEEL INSTITUTE, 150 E. 42nd St., New York, N.Y. 10017. Members are manufacturers and individuals in the basic steel industry. Publication:

Steel Facts

ASBESTOS TEXTILE INSTITUTE, P.O. Box 471, 131 N. York Rd., Willow Grove, Pennsylvania 19090. Organization of manufacturers of asbestos textiles and insulation fabrics. Publication:

Asbestos, monthly

THE ASPHALT INSTITUTE, Asphalt Institute Bldg., College Park, Maryland 20740. Tel.: (301) 927-0422. Association of refiners of asphalt from crude petroleum. Conducts programs of research and education. Publication:

Asphalt, quarterly

ASPHALT ROOFING MANUFACTURERS ASSOCIATION, 757 Third Ave., New York, N.Y. 10017. Tel.: (212) 421-2690.

BARRE GRANITE ASSOCIATION, P.O. Box 481, Barre, Vermont 05641.

COPPER DEVELOPMENT ASSOCIATION, 405 Lexington Ave., New York, N.Y. 10017. Tel.: (212) 867-6500. Group of more than 75 companies in mining, smelting and refining, and fabrication. Offers technical support for both the industry and its customers.

FLAT GLASS MARKETING ASSOCIATION, 1325 Topeka Ave., Topeka, Kansas 66612. Tel.: (913) 232-8231. Organization of glass distributors and contractors assists architects and engineers in the use of plate and window glass in buildings.

LEAD INDUSTRIES ASSOCIATION, 292 Madison Ave., New York, N.Y. 10017. Tel.: (212) 679-6020. Offers data and information on lead and lead products. Maintains an extensive library. Publication:

Lead, semiannual

NATIONAL CELLULOSE INSULATION MANUFACTURERS ASSOCIATION, Box 1241, Lima, Ohio 45802.

NATIONAL MINERAL WOOL INSULATION ASSOCIATION, 211 E. 51st St., New York, N.Y. 10022. Tel.: (212) 758-5210.

PERLITE INSTITUTE, 45 W. 45th St., New York, N.Y. 10036. Tel.: (212) 265-2145. Organization of processors of volcanic ash used for insulation and as an aggregate in plaster and concrete.

ROOF DRAINAGE MANUFACTURERS INSTITUTE, 221 N. LaSalle St., Chicago, Illinois 60601. Tel.: (312) 332-7127. Organization of manufacturers of gutters and leaders.

SEALED INSULATING GLASS MANUFACTURERS ASSOCIATION, Suite 209, 202 S. Cook St., Barrington, Illinois 60010. Tel.: (312) 381-8989.

THERMAL INSULATION MANUFACTURERS ASSOCIATION, 7 Kirby Plaza, Mt. Kisco, N.Y. 10549. Tel.: (914) 241-2284.

VERMICULITE ASSOCIATION, 52 Executive Park South, Atlanta, Georgia 30329. Tel.: (404) 631-5621. Organization of manufacturers fosters the use of vermiculite in lightweight concrete and for acoustic and thermal insulation.

ZINC INSTITUTE, 292 Madison Ave., New York, N.Y. 10017. Tel.: (212) 679-6020. Offers information to anyone interested in the utilization of zinc and zinc-coated products.

Periodicals

ARCHITECTS' GUIDE TO GLASS, METAL, & GLAZING, official publication of the Flat Glass Marketing Association. Published annually by U.S. Glass Publications, 2158 Union Ave., Suite 401, Memphis, Tennessee 38104. Covers sealants, glazing, flat glass, storefronts, curtain walls, window walls, doors, entrances, hardware, mirrors, etc. $3 per copy.

GLASS DIGEST, published monthly by Ashlee Publishing Co., 15 E. 40th St., New York, N.Y. 10016. Covers retailing, wholesaling, and contract glazing involving glass, storefront construction, curtain wall erection, doors and entrances, windows, hardware, building specialties, mirrors, etc. $8 per year.

U.S. GLASS, METAL, AND GLAZING, official publication of the Flat Glass Marketing Association. Published six times a year by U.S. Glass Publications, 2158 Union Ave., Suite 401, Memphis, Tennessee 38104. For wholesalers, retailers, and installers. No cost to qualified readers; $6 per year for others. Buyers' Guide published annually.

Directory

GLASS/METAL CATALOG, published annually by Artlee Catalog Inc., 15 E. 40th St., New York, N.Y. 10016. Standard reference of the flat glass, architectural metal, and allied products industry. Contains directories of suppliers, resources, brand names, and supplier literature, and a cross-referenced index. No cost to qualified readers, $10 per copy for others.

IV. CONTRACTORS, BUILDERS, CONSTRUCTION MANAGERS

This section covers the doers in construction—the contractors and subcontractors who can handle anything from excavation to wall coverings. Contractor associations are listed. A particularly rich selection of periodicals is provided, including many state and regional contractor publications. And finally, directories related to contracting are catalogued.

Associations and Professional Societies

ALLIED CONSTRUCTION EMPLOYERS ASSOCIATION, 2733 W. Wisconsin Ave., Milwaukee, Wisconsin 53208. Tel.: (414) 933-8110. Represents almost 1,000 contractors in the Milwaukee area. Conducts programs in construction safety, management, education, and apprentice training.

AMERICAN BUILDING CONTRACTORS ASSOCIATION, 3345 Wilshire Blvd., Los Angeles, California 90010. Tel.: (213) 385-0148. California organization of more than 2,500 general and specialty contractors and others involved with building. Publication:

American Building Contractor

AMERICAN INSTITUTE OF CONSTRUCTORS, Suite 511, 1140 N.W. 63rd St., Oklahoma City, Oklahoma 73116. Tel.: (405) 843-5531. Publication:

The Professional Constructor, annual

AMERICAN SOCIETY OF CONCRETE CONSTRUCTORS, 2510 Dempster St., Des Plaines, Illinois 60016. Tel.: (312) 296-7370. Publication:

Concrete Constructor, bimonthly

AMERICAN SUBCONTRACTORS ASSOCIATION, 806 15th St., N.W., Washington, D.C. 20005. Tel.: (202) 783-1883. Organization deals with problems of bid shopping, slow pay, and other contractual inequities. Publications:

Newsletter, monthly
Chapter Directory, annual

ASSOCIATED BUILDERS AND CONTRACTORS, Box 698, Glen Burnie, Maryland 21061. Tel.: (301) 760-6060. Association of contractors and subcontractors. Publications:

The Contractor, monthly
Fortnighter, quarterly
National Membership Directory

ASSOCIATED GENERAL CONTRACTORS OF AMERICA, 1957 E. St., N.W., Washington, D.C. 20006. Tel.: (202) 393-2040. Organization of general contractors engaged in commercial, industrial, and public utility construction. Publication:

Constructor, monthly

ASSOCIATED PUBLIC WORKS CONTRACTORS, 2835 N. Mayfair Rd., Milwaukee, Wisconsin 53222. Tel.: (414) 788-1050.

BUILDING MAINTENANCE CONTRACTORS ASSOCIATION, 2017 Walnut St., Philadelphia, Pennsylvania 19103. Tel.: (215) 569-3650. Publication:

The Bulletin, monthly

BUILDING WATERPROOFERS ASSOCIATION, 60 E. 42nd St., New York, N.Y. 10017. Tel.: (212) 682-4811.

CEILINGS AND INTERIOR SYSTEMS CONSTRUCTORS ASSOCIATION, 1201 Waukegan Rd., Glenview, Illinois 60025. Tel.: (312) 724-7700. Publication:

Sound Ideas, bimonthly

CONSTRUCTION INDUSTRY EMPLOYERS ASSOCIATION, 361 Delaware Ave., Buffalo, N.Y. 14202. Tel.: (716) 854-2340.

COUNCIL OF MECHANICAL SPECIALTY CONTRACT-

ING INDUSTRIES, 7315 Wisconsin Ave., Bethesda, Maryland 20014. Tel.: (301) 657-3110. Organization of almost 7,000 electrical, air-conditioning, ventilation, and plumbing contractors. Publication:

Councilator, monthly

ELECTRICAL CONTRACTORS ASSOCIATION, 808 N. Third St., Suite 406, Milwaukee, Wisconsin 53203. Tel.: (414) 273-6916.

GENERAL BUILDING CONTRACTORS ASSOCIATION, Suite 1212, 2 Penn Center Plaza, Philadelphia, Pennsylvania 19102. Tel.: (215) 568-7015.

GYPSUM DRYWALL CONTRACTORS INTERNATIONAL, Suite 600, 2010 Massachusetts Ave., N.W., Washington, D.C. 20036. Tel.: (202) 872-0100. Publication:

Drywall Newsmagazine, bimonthly

GYPSUM ROOF DECK FOUNDATION, 5820 N. Nagle Ave., Chicago, Illinois 60646. National association of gypsum roof deck contractors. Publication:

GRDF Newsletter

INTERNATIONAL ASSOCIATION OF WALL AND CEILING CONTRACTORS, 20 E. St., N.W., Washington, D.C. 20001. Tel.: (202) 638-1072. Publications:

Technical Topics Digest, quarterly

Walls and Ceilings, monthly

INTERNATIONAL MASONRY INSTITUTE, 823 15th St., N.W., Suite 1001, Washington, D.C. 20005. Tel.: (202) 783-3908.

LAND IMPROVEMENT CONTRACTORS OF AMERICA, 410 N. Michigan Ave., Chicago, Illinois 60611. Tel.: (312) 321-1470. Publication:

Land and Water Development Magazine, monthly

MASON CONTRACTORS ASSOCIATION OF AMERICA, 208 S. LaSalle St., Chicago, Illinois 60604. Tel.: (312) 726-5742. Publication:

Masonry

MECHANICAL CONTRACTORS ASSOCIATION OF AMERICA, 5530 Wisconsin Ave., N.W., Suite 750, Washington, D.C. 20015. Tel.: (202) 654-7960. Organization of piping contractors for heating, refrigerating, ventilation, and air conditioning. Publications:
Government and Labor Reporter, monthly
The Reporter, monthly

NATIONAL ASSOCIATION OF BUILDING SERVICE CONTRACTORS, 2215 M St., N.W., Washington, D.C. 20037. Tel.: (202) 659-1821.

NATIONAL ASSOCIATION OF ELEVATOR CONTRACTORS, 4321 Hartwick Rd., College Park, Maryland 20740. Tel.: (301) 927-3338. Group of contractors and manufacturers who install and supply elevator equipment. Has developed elevator standards for architects and others.

NATIONAL ASSOCIATION OF FLOOR COVERING INSTALLERS, 4301 Connecticut Ave., N.W., Suite 436, Washington, D.C. 20008. Tel.: (202) 966-2082. Carpet and resilient floor contractors' organization. Publications:
NAFCI News, monthly
Membership Directory, annual

NATIONAL ASSOCIATION OF HOME BUILDERS, 1625 L St., N.W., Washington, D.C. 20036. Tel.: (202) 737-7435. Organization of home and apartment builders, architects, contractors, and others, with membership of about 70,000 in 33 states and 490 chapters. Publications:
NAHB Journal-Scope, weekly
Compendium of Multifamily Housing
Economic News Notes, monthly

NATIONAL ASSOCIATION OF LIGHTING MAINTENANCE CONTRACTORS, 3725-C N. 126th St., Brookfield, Wisconsin 53005. Tel.: (414) 781-2028. Publication:
NALMCO Main-Lighter, bimonthly

NATIONAL ASSOCIATION OF MINORITY CONTRACTORS, World Trade Center, San Francisco, California 94111. Tel.: (415) 398-0484.

NATIONAL ASSOCIATION OF MISCELLANEOUS, ORNAMENTAL, AND ARCHITECTURAL PRODUCTS CONTRACTORS, Suite 300, 10533 Main St., Fairfax, Virginia 22030. Tel.: (703) 591-6485. Publications:
Newsletter, monthly
Membership Roster

NATIONAL ASSOCIATION OF PLUMBING-HEATING-COOLING CONTRACTORS, 1016 20th St., N.W., Washington, D.C. 20036. Tel.: (202) 337-1675. Group of more than 400 local plumbing, heating, and cooling contractor organizations with a total membership of about 9,000.

NATIONAL ASSOCIATION OF REINFORCING STEEL CONTRACTORS, Suite 300, 10533 Main St., Fairfax, Virginia 22030. Tel.: (703) 591-6485. Organization of contractors who place reinforcing steel and post-tensioning systems. Publications:
Newsletter, monthly
Membership Roster

NATIONAL ASSOCIATION OF RIVER AND HARBOR CONTRACTORS, Suite 536, Washington Bldg., Washington, D.C. 20005. Tel.: (202) 783-2470. The 30 companies that compose the association are engaged in dredging and landfill operations, and in waterways development.

NATIONAL CONSTRUCTORS ASSOCIATION, 1133 15th St., N.W., Washington, D.C. 20005. Tel.: (202) 466-8880. Publication:
NCA Newsletter, monthly

NATIONAL ELECTRICAL CONTRACTORS ASSOCIATION, 7315 Wisconsin Ave., 13th Floor, Washington, D.C. 20014. Tel.: (301) 657-3110. Members are electrical contractors who install, repair, and maintain electric wiring and equipment. Conducts management and training programs for electrical contractors. Publications:
Electrical Contractor
NECA Newsletter, weekly

NATIONAL ENVIRONMENTAL SYSTEMS CONTRACTORS ASSOCIATION, 1501 Wilson Blvd., Arlington, Virginia

22209. Composed of contractors who install air conditioning, refrigerating, and heating equipment, and manufacturers who produce the equipment. Publication:
NESCA News, monthly

NATIONAL HOME IMPROVEMENT COUNCIL, 11 E. 44th St., New York, N.Y. 10017. Tel.: (212) 867-0121. Council includes members representing contractors, manufacturers, and others from the home improvement industry. Publication:
NHIC Newsletter

NATIONAL INSULATION CONTRACTORS ASSOCIATION, 8630 Fenton St., Suite 506, Silver Spring, Maryland 20910. Tel.: (301) 585-6623. Composed of contractors and suppliers of insulation. Publications:
NICA Outlook, monthly
NICA Insulator, bimonthly
NICA Roster, annual

NATIONAL REMODELERS ASSOCIATION (NERSICA), 50 E. 42nd St., New York, N.Y. 10017. Tel.: (212) 687-5224. More than 5,000 members involved in home improvements and industrial remodeling, and in the manufacture of home improvement products. Publication:
Remodeling Contractor, monthly

NATIONAL ROOFING CONTRACTORS ASSOCIATION, 1515 N. Harlem Ave., Oak Park, Illinois 60302. Tel.: (312) 383-9513. Organization of the waterproofing, roofing, siding, and insulating contracting business.

NATIONAL STEEL ERECTORS ASSOCIATION, Suite 904, 1800 N. Kent St., Arlington, Virginia 22209. Tel.: (703) 524-3333.

NATIONAL UTILITY CONTRACTORS ASSOCIATION, 815 15th St., N.W., Washington, D.C. 20005. Tel.: (202) 737-2133. Publication:
Directory of Members, annual

PAINTING AND DECORATING CONTRACTORS OF AMERICA, 2625 W. Peterson Ave., Chicago, Illinois 60659. Tel.: (312) 561-2328. Publications: *PDCA,* monthly

SHEET METAL AND AIR CONDITIONING CONTRACTORS NATIONAL ASSOCIATION, 1611 N. Kent St., Arlington, Virginia 22209. Tel.: (703) 337-3380. Publication:
Newsletter, biweekly

SOCIETY OF CONSTRUCTION SUPERINTENDENTS, 150 Nassau St., New York, N.Y. 10038.

TILE CONTRACTORS ASSOCIATION OF AMERICA, 112 N. Alfred St., Alexandria, Virginia 22314. Tel.: (703) 836-5995. Organization of 450 ceramic tile contractors. Publication:
Tile and Decorative Surfaces, bimonthly

UNDERGROUND CONTRACTORS ASSOCIATION, 8550 W. Bryn Mawr Ave., Chicago, Illinois 60631. Tel.: (312) 693-6930.

Periodicals

AIR CONDITIONING AND REFRIGERATION BUSINESS, published monthly by the Industrial Publishing Co., 614 Superior Ave., West Cleveland, Ohio 44113. For those in the fields of air conditioning, commercial and industrial refrigeration, heating, and air handling. No cost to qualified readers; $12 per year for others.

ALABAMA CONTRACTOR, published monthly by Associated Plumbing, Heating, and Cooling Contractors of Alabama, 2231 S. 20th Ave., Birmingham, Alabama 35223. No cost to qualified readers; $2 per year for others.

ALASKA CONSTRUCTION AND OIL REPORT, published monthly by Alaska Construction News Inc., 109 W. Mercer, Seattle, Washington 98119. Covers current and planned projects, material and equipment developments, bid invitations and results, contractors' unit prices, personnel, and technical data. No cost to qualified readers; $10 per year for others.

AMERICAN ARTISAN, published monthly by Plumbing and Heating Publishing Co., 317 Howard St., Evanston, Illinois 60202. For dealer-contractors who sell, fabricate, install, and service air conditioning systems (winter and summer) for residential, commercial, and industrial buildings, and for sheet metal contractors who fabricate and install architectural sheet metal and air handling systems for industrial ventilation, dust removal, large air conditioning systems, and warm air heating. No cost to qualified readers; $5 per year for others.

AMERICAN PAINTING CONTRACTOR, published monthly by American Paint Journal Co., 2911 Washington Ave., St. Louis, Missouri 63103. No cost to qualified readers; $5 per year for others.

AMERICAN ROOFER & BUILDING IMPROVEMENT CONTRACTOR, published monthly by Shelter Publications, 221 Lake St., Oak Park, Illinois 60302. For contracting firms installing roofing, siding, thermal insulation, and building improvements involving bitumens; asphalt, tarred asbestos, and glass fiber felts; brushed, sprayed, or sheet elastomeric materials; shingles of asphalt, asbestos cement, aluminum, or wood; roofing and siding of rigid vinyl and other plastics; sidings of asphalt, asbestos cement, aluminum, or plastic; storm doors and windows; porch railings and enclosures. $5 per year.

APARTMENT CONSTRUCTION NEWS, published monthly by Gralla Publications, 1301 Broadway, New York, N.Y. 10036. For builders, developers, architects, and contractors involved in multifamily housing. Covers site selection, building and design and layout, construction methods, materials, zoning, and financing. $3 per year.

ARIZONA-NEW MEXICO CONTRACTOR & ENGINEER, published monthly by Graphics West Inc., 806 N. 4th St., P.O. Box 3407; Phoenix, Arizona 85030. For contractors, engineers, subcontractors, equipment distributors, material suppliers, trade associations, and public officials in the heavy construction and public works fields. No cost to qualified readers; $4 per year for others.

BUILDER, official publication of Associated General Contractors of Illinois. Published monthly by Builder Magazine, Inc., 3219 Executive Park Dr., P.O. Box 2579, Springfield, Illinois 62708. For those in the heavy, highway, and utility construction industries responsible for planning, construction, and maintenance of highways and other public works. $4 per year.

BUILDERS JOURNAL, official publication of the Building Industry Association of California. Published monthly by Frank M. McKellar and Associates, 1830 W. 8th St., Los Angeles, California 90057. For state-licensed builders, general contractors, and developers in California, Arizona, Nevada, and New Mexico. Covers single and multifamily buildings, remodeling, churches, schools, commercial and industrial construction, and public works. No cost to qualified readers; $6 per year for others.

BUILDERS REPORT PACIFIC, published weekly by the Builders Report Service, 740 Ala Moana, Honolulu, Hawaii 96813. For contractors, subcontractors, architects, engineers, and builder-developers in the state of Hawaii. Covers jobs out for bid, bid results, contract awards, jobs contemplated or being planned, new business licenses issued, building permits issued, new products, and improved methods of construction.

BUILDING, published weekly by Building (Publishers) Ltd., The Builder House, P.O. Box 135, 4 Catherine St., London WC2B 5JN, England. For architects, quantity surveyors, construction engineers, contractors, and clients. Outside U.K., £9 per year.

BUILDING NEWS, published weekly at 3055 Overland Ave., Los Angeles, California 90034. No cost to qualified readers; $6 per year for others.

BUILT ENVIRONMENT, published by Building (Publishers) Ltd., The Builder House, P.O. Box 135, 4 Catherine St., London WC2B 5JN, England. For planners, architects, and engineers. Outside U.K., £6.50 per year.

CALIFORNIA BUILDER, official publication of Associated Home Builders, San Francisco, Marin, Sonoma, and Napa Counties; Associated Home Builders of Greater Eastbay; Builders Association of Santa Clara-Santa Cruz Counties; Peninsula Building Industry Association; Home Builders Association of Contra Costa County; Home Builders Association of Central California; Builders Association of Northern Nevada. Published monthly by Fellon Publishing Co., 693 Mission St., Penthouse, San Francisco, California 94105. For entrepreneur builders, developers, and licensed general building contractors engaged in residential building. No cost to qualified readers; $10 per year for others.

CALIFORNIA BUILDER & ENGINEER, published semimonthly by California Builder & Engineer Inc., 363 El Camino Real, South San Francisco, California 94080. For the heavy construction industry of California. No cost to qualified readers; $12.50 per year for others.

CEE—CONTRACTORS' ELECTRICAL EQUIPMENT, published monthly by Sutton Publishing Co., 172 Broadway, White Plains, New York 10605. For electrical contractors, wholesalers, engineers, architects, electrical utilities, and modular and mobile building manufacturers. No cost to qualified readers.

CENTRAL STATES CONSTRUCTION MAGAZINE, published monthly by Kansas Construction Magazine Inc., 4125 Gage Center Dr., Topeka, Kansas 66604. For those who specify or buy construction equipment, materials, and supplies in the central states. No cost to qualified readers; $10 per year for others.

COMMERCE BUSINESS DAILY, published daily (Monday through Friday) by the Government Printing Office, Washington, D.C. 20402. Synopsis of U.S. Government proposed procurement, sales, and contract awards. $40 per year.

CONSTRUCTION, published every other Monday by Construction Pub. Co., 2420 Wilson Blvd., Arlington, Virginia 22201. For contractors, material producers, architects, engineers, equipment manufacturers, distributors, and government officials concerned with design, construction, maintenance, and administration of highways, heavy construction, public works, and industrial and nonresidential buildings in Maryland, District of Columbia, Virginia, West Virginia, and North Carolina. No cost to qualified readers; $20 per year for others.

CONSTRUCTION BARGAINEER, published twice every month by Construction Bargaineer, P.O. Box 1061, St. Paul, Minnesota 55105. Coverage includes buildings, roads, tunnels, dams, bridges, pipelines, irrigation, logging, railroads, mining, quarrying, ships, marinas, government surplus, excavating, drilling, dredging, heavy hauling, house moving, landscaping, and chemical and fertilizer plants. $5 per year.

CONSTRUCTION BULLETIN, published weekly by Chaping Publishing Co., 1022 Upper Midwest Bldg., Minneapolis, Minnesota 55401. Provides current and advance information on construction of highways, airports, bridges, sewers, waterworks, and heavy public and private buildings in Minnesota, North Dakota, and South Dakota. $19.50 per year.

CONSTRUCTION DIGEST, published biweekly by Construction Digest Inc., 101 E. 14th St., P.O. Box 603, Indianapolis, Indiana 46206. For the construction and public works industries in Illinois, Indiana, Ohio, Kentucky, and eastern Missouri, and for the aggregate-producing and mining fields and users of heavy equipment in industry and utilities. Two editions: East and West. No cost to qualified readers; for others, $15 per year (both editions, $28).

CONSTRUCTIONEER, published monthly by Reports Corp., 1 Bond St., Chatham, N.J. 07928. For contractors, public officials, producers, and suppliers in the states of Pennsylvania, New York, New Jersey, and Delaware. Covers bids called, low bids, awards, and plans for roads, bridges, streets, airports, sanitation facilities, federal projects, housing, public buildings, commercial and industrial installations, and parks. No cost to qualified readers; $20 per year for others.

CONSTRUCTION EQUIPMENT, published 13 times a year by Conover-Mast Publications, 205 E. 42nd St., New York, N.Y. 10017. Reports on trends in the construction market, field applications, equipment and management news, and new products. No cost to qualified readers, $20 per year for others. Buyers Guide published annually in April.

CONSTRUCTION MACHINERY MAINTENANCE, published bimonthly by Abrahamson Publishing Co., P.O. Box 190, Barrington, Illinois 60010. Covers tools, components, and methods. No cost to qualified readers.

CONSTRUCTION METHODS & EQUIPMENT, published monthly by McGraw-Hill Publications, 1221 Avenue of the

Americas, New York, N.Y. 10020. Covers techniques, equipment, and materials and their applications in building dams, highways, bridges, tunnels, railroads, airports, pipelines, water supply and sewage systems, buildings, industrial plants, etc. $4 per year.

CONSTRUCTION NEWS, published every other week by Construction News Inc., 715 W. Second St., P.O. Box 2421, Little Rock, Arkansas 72203. Covers news of the construction industry (exclusive of small home building) in the states of Oklahoma, Arkansas, western Tennessee, Mississippi, and Louisiana. No cost to qualified readers; $25 per year for others.

CONSTRUCTION REPORTS, published by the Government Printing Office, Washington, D.C. 20406. HOUSING STARTS, monthly. Order No. C56.211/3, $4.50 per year. VALUE OF CONSTRUCTION PUT IN PLACE, monthly. Order No. C56.211/5, $5.50 per year. HOUSING AUTHORIZED BY BUILDING PERMITS AND PUBLIC CONTRACTS, monthly. Order No. C56.211/4, $17.50 per year.

CONSTRUCTION REVIEW, published monthly by the Government Printing Office, Washington, D.C. 20406. $11.50 per year.

CONTRACTOR, published twice a month by Buttenheim Publishing Corp., Berkshire Common, Pittsfield, Massachusetts 01201. For wholesalers and contractors in the plumbing, heating, and air conditioning industry. No cost to qualified readers; $15 per year for others.

CONTRACTORS & ENGINEERS MAGAZINE, published monthly by Buttenheim Publishing Corp., Berkshire Common, Pittsfield, Massachusetts 01201. Covers highways, heavy building, and other heavy construction, including field operations, management, equipment maintenance, and new products. No cost to qualified readers; $15 per year for others.

CONTRACTORS EQUIPMENT LOCATOR, published monthly by Key Information Systems Inc., 419 7th St., N.W., Washington, D.C. 20004. Descriptive classified listings of heavy construction equipment for sale or lease, wanted, and reported stolen. No cost to qualified readers; $12.50 for others.

CURRENT HOUSING REPORTS, published by the Government Printing Office, Washington, D.C. 20402. HOUSING VACANCIES, quarterly, and HOUSING CHARACTERISTICS, published occasionally. $3.50 annually for both.

DAILY CONSTRUCTION SERVICE, LOS ANGELES EDITION, published daily by Withers Publishing Co., 443 S. Hill St., Suite 818, Los Angeles, California 90013. For those in the construction industry responsible for building dams, sewers, highways, airports, bridges, harbors, and major public buildings. Covers bid calls, results, and contracts awarded. $200 per year.

DAILY CONSTRUCTION SERVICE, SAN FRANCISCO EDITION, published daily by Wade Publishing Co., P.O. Box 3019, 230 California St., Suite 400, San Francisco, California 94119. See Los Angeles Edition, above.

DAILY JOURNAL, published weekdays by McGraw-Hill Information Systems Co., 1217 Welton St., Denver, Colorado 80204. Covers construction, law, real estate, insurance, and finance in Colorado and Wyoming. Information on bidders, low bidders, awards, building permits, and plans on file. Reports from federal, state, and local courts on judgments, bankruptcies, real estate transfers, chattel mortgages, liens, and new suits. $240 per year.

DAILY JOURNAL OF COMMERCE OF PORTLAND, OREGON, published mornings except Sundays by Journal of Commerce, 2014 N.W. 24th., Portland, Oregon 97210. $0.15 per copy.

DAILY PACIFIC BUILDER, published weekdays by McGraw-Hill Information Systems Co., 2450 17th St., San Francisco, California 94110. Construction management publication for architects, engineers, general contractors, subcontractors, and suppliers. $300 per year.

DE/JOURNAL (Domestic Engineering Journal), published monthly by Construction Industry Press, 522 N. State Rd., Briarcliff Manor, N.Y. 10510. For firms engaged in plumbing, heating, air conditioning, process piping, and water systems construction. In three editions: (1) residential work and water systems work regardless of size, (2) commercial, industrial, and institutional work including high rise, (3) editions 1 and 2 combined. No cost to qualified readers; $10 for others. Special issue: Book of Giants, listing the 200 largest contractors including key personnel, breakdown of activity, descriptions of recent jobs. Issued annually in August.

DIXIE CONTRACTOR, published biweekly by Dixie Contractor, 525 Marshall St., P.O. Box 280, Decatur, Georgia 30031. For contractors, material producers, federal, state, and city public officials, distributors, and manufacturers of the construction industry in the Southeast. No cost to qualified readers; $7.50 for others.

DODGE CONSTRUCTION NEWS, CHICAGO EDITION, published weekdays by McGraw-Hill Information Systems Co., 230 W. Monroe St., Chicago, Illinois 60606. Daily review of construction projects being bid, awaiting awards, and awarded, and general news of interest to the construction industry. No cost to qualified readers; $334 per year for others.

EARTHMOVING & CONSTRUCTION, published monthly by Eastern Publications, 76 Mt. Kemble Ave., Morristown, N.J. 07960. Covers coal stripping, road and dam building, excavating, quarrying, sand and gravel pits, pipe laying, and other construction involving mass earthmoving. $10 per year.

ELECTRIC HEAT, published bimonthly by Barks Publications, 360 N. Michigan Ave., Chicago, Illinois 60601. For businesses distributing, selling, and installing electric heating for residential, commercial, and industrial buildings. No cost for qualified readers; $10 per year for others.

ELECTRICAL CONTRACTOR, published monthly by National Electrical Contractors Association, 7315 Wisconsin Ave., Washington, D.C. 20014. For electrical contractors, electrical consulting engineers, wholesalers, inspectors, and utility executives. No cost to qualified readers; $10 per year for others.

ENGINEERING NEWS-RECORD, published weekly by McGraw-Hill Publications, 1221 Avenue of the Americas, New

York, N.Y. 10020. Covers legislation, labor, finance, prices, and trends and developments in design and construction concerned with buildings, transportation, and water resources. $9 per year.

EQUIPMENT GUIDE NEWS, published monthly by Equipment Guide-Book Co., 3980 Fabian Way, Palo Alto, California 94303. For those who select and purchase equipment in the heavy construction industry. No cost to qualified readers; $2 per year for others.

EXCAVATING CONTRACTOR, published monthly by Cummins Publishing Co., 21590 Greenfield Rd., Oak Park, Michigan 48237. For smaller contractors engaged in earthmoving, including excavating for foundations, pipelines, ditches, trenches, sewers, septic systems, site preparation, and backfilling. No cost to qualified readers; $10 per year for others.

FARM BUILDING NEWS, published six times a year by American Farm Building Services, 733 N. Van Buren, Milwaukee, Wisconsin 53202. For farm building contractors. No cost to qualified readers; $5 per year for others.

FLORIDA CONTRACTOR, published monthly by Wesley Associates, P.O. Box 510, Lakeland, Florida 33802. Official publication of Associated Plumbing and Mechanical Contractors of Florida. For plumbing, heating, and cooling contractors, wholesalers, and plumbing inspectors. No cost to qualified readers; $4 per year for others.

FLORIDA BUILDER, published monthly by Peninsula Publishing Co., 2306 S. Hubert Ave., Tampa, Florida 33609. Ideas, products, methods, and materials for the residential and commercial building fields. No cost to qualified readers; $3 per year for others.

FLORIDA CONTRACTOR AND BUILDER, published monthly by Hopkins Publication, 3620 N.W. 7th St., Miami, Florida 33125. Covers both heavy and light construction. No cost to qualified readers; $5 per year for others.

FUEL OIL & OIL HEAT, published monthly by Industry Publications, 200 Commerce Rd., Cedar Grove, New Jersey 07009.

For fuel oil and oil heating dealers and heating and air conditioning contractors. $7 per year.

HEATING & AIR CONDITIONING CONTRACTOR, published monthly by Edwin A. Scott Publishing Corp., 522 N. State Rd., Briarcliff Manor, N.Y. 10510. Covers sales, fabrication, installation, and servicing of warm air heating, central and packaged air conditioning, ventilation, roofing, and sheet metal work. No cost to qualified readers; $5 per year for others. Annual Buyers Guide.

HOME IMPROVEMENTS, published monthly by Peacock Business Press, 200 S. Prospect Ave., Park Ridge, Illinois 60068. Covers selling and marketing techniques, merchandising, lead development, cost estimating, business procedures, government regulations, profit management. No cost to qualified readers; $5 per year for others.

HOMEBUILDING BUSINESS, published monthly by Gralla Publications, 1501 Broadway, New York, N.Y. 10036. For builders and developers of single-family homes and communities. Covers business management, finance, marketing, merchandising, zoning and government relations, project planning, architecture, construction technology, and industry trends. $9 per year.

HOUSE & HOME, published monthly by McGraw-Hill Publications, 1221 Avenue of the Americas, New York, N.Y. 10020. Covers planning, architecture, building methods, land buying and development, finance, management, and marketing in the field of housing and other light construction. $6 per year.

ILLINOIS MASTER PLUMBER, published monthly by Illinois Association of Plumbing-Heating-Cooling Contractors, Marquette Bldg., Suite 273, 140 S. Dearborn St., Chicago, Illinois 60603. No cost to qualified readers; $2 per year for others.

INDIANA PLUMBING-HEATING-COOLING CONTRACTOR, published monthly by the Indiana P-H-C Contractor Inc., 2201 E. 46th St., Indianapolis, Indiana 46205. For contractors, wholesalers, specifying architects and engineers, builders, and building owners in Indiana. No cost to qualified readers; $5 per year for others.

IOWA PLUMBING, HEATING, COOLING CONTRACTOR, published monthly by the Iowa Association of Plumbing Contractors, IHSAA Bldg., P.O. Box 56, Boone, Iowa 50036. For those in the plumbing-heating-cooling contracting industry, and for state procurement authorities. No cost to qualified readers.

MASONRY, published 11 times a year by the Mason Contractors Association of America, 208 S. LaSalle St., Suite 480, Chicago, Illinois 60604. For mason contractors, brick and concrete masonry manufacturers and dealers, architects, engineers, and union officials. $5 per year.

MASONRY INDUSTRY, published monthly by Building News, 3055 Overland Rd., Los Angeles, California 90034. No cost to qualified readers; $4.25 per year for others.

METRO ATLANTA BUILDER, published bimonthly by John Yopp Publications, P.O. Box 7368, Atlanta, Georgia 30309. For those responsible for planning, designing, financing, developing, and erecting residential and commercial structures in the 10-county metropolitan Atlanta area. No cost to qualified readers; $5 per year for others.

MID-WEST CONTRACTOR, published biweekly by Mid-West Records Inc., P.O. Box T66, 2537 Madison Ave., Kansas City, Missouri 64141. For the construction industry in Iowa, Nebraska, Kansas, and western and northeastern Missouri. No cost to qualified readers; $15 per year for others. Directory issued annually.

OHIO CONTRACTOR, published monthly by Ohio Contractor, 1375 W. Lane Ave., Columbus, Ohio 43221. For Ohio's highway, heavy, municipal, utility, and building construction industries. No cost to qualified readers; $6.50 per year for others. Directory issued annually.

PACIFIC BUILDER & ENGINEER, published semimonthly by Publishers Professional Services, 109 W. Mercer St., Seattle, Washington 98119. For the heavy construction industry of Oregon, Washington, Idaho, and Montana. No cost to qualified readers; $7.50 per year for others. Buyers Guide and Directory issued annually.

PENNSYLVANIA CONTRACTOR, published monthly by the Pennsylvania Association of Plumbing Contractors, 219 Pine St., Harrisburg, Pennsylvania 17101. For plumbing, heating, and cooling contractors in Pennsylvania. No cost to qualified readers; $4 per year for others.

PLUMBING-HEATING-COOLING BUSINESS, published monthly by Plumbing and Heating Publishing Co., 317 Howard St., Evanston, Illinois 60202. For the contractor who sells and installs plumbing, heating, air conditioning, refrigeration, piping, and related mechanical systems. No cost to qualified readers; $10 per year for others.

PROFESSIONAL BUILDER & APARTMENT BUSINESS, published monthly by Cahners Publishing Co., 5 S. Wabash Ave., Chicago, Illinois 60603. Covers single-family homes, low-rise apartments, and light nonresidential buildings. No cost to qualified readers; $24 per year for others.

REMODELING CONTRACTORS, published monthly by Remodeling Contractor, 50 E. 42nd St., New York, N.Y. 10017. For home improvement and light commercial remodeling contractors. No cost to qualified readers; $10 per year for others.

ROCKY MOUNTAIN CONSTRUCTION, published biweekly by Mountain Publishing Co., 2201 Stout St., Denver, Colorado 80205. Covers heavy engineering, public and private building, landscaping, soil conservation, mining, logging, and federal, state, county, and city projects. Companion daily construction reports also published. No cost to qualified readers; $15 per year for others. Annual Buyers Guide and Directory.

RSC REFRIGERATION SERVICE AND CONTRACTING, official publication of Refrigeration Service Engineers Society. Published monthly by Nickerson and Collins Co., 2720 Des Plaines Ave., Des Plaines, Illinois 60013. Covers theory, techniques, and applications. $5 per year.

RSI (ROOFING, SIDING, AND INSULATION), published monthly by Harcourt Brace Jovanovich Inc., 757 Third Ave., New York, N.Y. 10017. For contractors and wholesalers of roofing, siding, insulation, and related products including sheet metal, rain-carrying equipment, ventilators, skylights, waterproofing and exterior coatings, and decking. No cost to qualified readers; $6 per year for others.

SERVICE REPORTER, published monthly by Technical Reporting Corp., P.O. Box 745, 1098 Milwaukee Rd., Wheeling, Illinois 60090. Covers sales, installation, and service of air conditioning, heating, and refrigeration equipment and appliances. No cost to qualified readers.

SNIPS MAGAZINE, published monthly by Snips Magazine Inc., 407 Mannheim Rd., Bellwood, Illinois 60104. For sheet metal, warm air heating, air conditioning, and ventilating contractors, manufacturers, and wholesalers. No cost to qualified readers; $2 per year for others.

SOUTHERN CALIFORNIA HEAVY CONSTRUCTION, published nine times a year by Earth Publications, 8402 Allport Ave., Drawer 2263, Santa Fe Springs, California 90670. Covers heavy construction in the 11 southernmost California counties. No cost to qualified readers.

SOUTHERN PLUMBING, HEATING, COOLING, published monthly by Southern Trade Publications, P.O. Box 9377, Greensboro, North Carolina 27408. For contractors and wholesalers in the plumbing, heating, and air conditioning fields in the 14 southern states from Maryland to Texas. $3 per year.

TEXAS CONTRACTOR, published semimonthly by Peters Publishing Co. of Texas, 2240 Vantage St., P.O. Box 1706, Dallas, Texas 75221. For contractors (engineering, building, and special trades), material producers, government officials, architects, engineers, equipment distributors and manufacturers. Covers plans in progress, bids wanted, low bidders, and contracts awarded. No cost to qualified readers; $18 per year for others.

UECA—UNDERGROUND ENGINEERING CONTRACTORS ASSOCIATION, published monthly by the Underground Engineering Contractors Association, 8615 Florence Ave., Suite 205, Downey, California 90240. For Southern California heavy construction contractors who specialize in storm drain, sewer, pipeline, waterworks, and tunnel construction, and for suppliers and public works officials.

WALLS & CEILINGS, official publication of the International Association of Wall and Ceiling Contractors. Published monthly by Plastering Industries Inc., Construction Center, 215 W. Harrison, Seattle, Washington 98119. For contractors, journeymen, building material dealers, and manufacturers of wall and ceiling products. No cost to qualified readers; $3 per year for others.

WESTERN BUILDER, published weekly by Western Builder Publishing Co., 6526 River Parkway, Milwaukee, Wisconsin 53213. Construction news service for contractors, subcontractors, material producers, public officials, architects, engineers, equipment manufacturers and distributors in Wisconsin, upper Michigan, and northern Illinois. $18 per year.

WESTERN CONSTRUCTION, published monthly by Westamerica Publications, 609 Mission St., San Francisco, California 94105. For heavy construction contractors, administrators, superintendents, engineers, public works officials, and equipment maintenance supervisors in the 13 western states. No cost to qualified readers; $10 per year for others.

WESTERN PLASTERER, published monthly by California Lumber Merchant, 573 S. Lake Ave., Pasadena, California 91101. Official publication of the California Lathing and Plastering Contractors Association. For construction contractors using lathing and plastering materials, systems, and equipment in California, Arizona, and Nevada. No cost to qualified readers; $4 per year for others.

WRECKING & SALVAGE JOURNAL, published monthly by the Wrecking and Salvage Journal Inc., P.O. Box 130, Hingham, Massachusetts 02043. For people engaged in building demolition, urban renewal, and salvage operations. No cost to qualified readers; $15 per year for others.

Directories

ABC DIRECTORY, published annually by ABC Publishing Co., 4128 College Ave., Des Moines, Iowa 50301. Products and services listed by city. Also lists architectural firms, engineering firms, and members of the Associated General Contractors and Master Builders of Iowa, public officials, commissions, and boards. $25 per copy.

AGC DIRECTORY ISSUE, CONSTRUCTOR magazine, published by the Associated General Contractors of America, 1957 E St., N.W., Washington, D.C. 20006. Lists 9,000 member firms.

ANNUAL REPORT OF HOUSING'S GIANTS, PROFESSIONAL BUILDER magazine, published by Cahners Publishing Co., 5 S. Wabash Ave., Chicago, Illinois 60603. Lists housing construction firms that grossed more than $10 million in the previous year.

BLUE BOOK CONTRACTOR'S REGISTER, published by Contractor's Register Inc., Elmsford, N.Y. 10523. Listing of architects, contractors, and sources of construction materials.

BLUE BOOK OF MAJOR HOMEBUILDERS, published at 1559 Eton Way, Crofton, Maryland 21113. Descriptions of major home builders, home manufacturers, mobile home manufacturers, and new town community developers. $69.50.

BUILDERS COMMERCIAL AGENCY CREDIT REFERENCE BOOK. Credit ratings of contracting firms in the Chicago area. Builders Commercial Agency, 105 N. Oak Park Ave., Oak Park, Illinois.

BUILDING CONSTRUCTION EMPLOYERS DIRECTORY. Lists building employers' associations and building unions in the Chicago area. Published by Building Construction Employers Association of Chicago, 228 N. LaSalle St., Chicago, Illinois 60601.

BUILDERS DIRECTORY AND GUIDE. Architects, engineers, contractors, and building materials manufacturers in the construction industry in Alaska, Idaho, Montana, Oregon, and Washington are listed. Published annually by Washington-Alaska Publishing Co., 120 W. Spokane St., Seattle, Washington 98134.

BUILDERS INSTITUTE OF WESTCHESTER AND PUT-

NAM COUNTIES YEARBOOK AND BUYERS GUIDE. Compilation of building contractors and building material suppliers. Updated annually. Builders Institute of Westchester and Putnam Counties, 1 E. Post Rd., White Plains, N.Y. 10601. No charge to members; others, $10.

BUILDING CONSTRUCTION INFORMATION SOURCES. How to get information on nearly 1,000 building topics. Gale Research Co., Book Tower, Detroit, Michigan 40226. $14.50

CONTRACTORS REGISTER. Published in two editions: (1) New York, New Jersey, Connecticut and (2) Washington, Baltimore, Philadelphia. Contractors Register Inc., 5 Van Wart St., Elmsford, N.Y. 10523. No cost to qualified readers.

DIRECTORY OF ELECTRICAL CONTRACTORS IN CALIFORNIA. Licensed firms and principals are listed with names, addresses, and phone numbers. W. Burrwood Jones Services, P.O. Box 11, Los Angeles, California 90053. $20.

DIRECTORY OF NATIONAL ELECTRICAL CONTRACTORS ASSOCIATION MEMBERS, published in *Electrical Contractor*. National Electrical Contractors Association, 7315 Wisconsin Ave., 13th Floor, Washington, D.C. 20014.

OFFICIAL DIRECTORY OF LICENSED CONTRACTORS. Names, addresses, and size descriptions of almost 10,000 licensed contractors in California. State of California Department of Professional and Vocational Standards, Contractors State License Board, 1020 N St., Sacramento, California 95814. $20.

EGCA DIRECTORY, published annually by Engineering and Grading Contractors Association, 8402 Allport Ave., P.O. Box 2263, Santa Fe Springs, California 90670. Lists construction equipment owned by EGCA by machinery classification and geographic location. A guide to construction equipment availability in California. No cost to qualified readers.

ILLINOIS CONTRACTORS TRADE DIRECTORY, official publication of Building Construction Contractors and Subcontracting Industry. Published by Illinois Contractors Trade Directory Inc., 220 S. State St., Chicago, Illinois 60604. No cost to qualified readers.

NATIONAL ROOFING CONTRACTORS ASSOCIATION DIRECTORY, published by National Roofing Contractors Association, 1515 Harlem Ave., Oak Park, Illinois 60302. Listing of officers, directors, and member firms, alphabetically and geographically.

SOURCE BOOK OF STATISTICS RELATING TO CONSTRUCTION. A compilation of annual, monthly, and quarterly construction data from government and private sources. Includes building permits, housing starts, construction contracts, and materials. National Bureau of Economic Research. $12.50.

V. STANDARDS, CODES, TESTING, INSPECTION, GOVERNMENT PLANNING

The extensive involvement of government in construction—at the local, state, and national levels—is reflected in the contents of this section. Here will be found governmental and quasigovernmental agencies responsible for planning, redevelopment, and building code enforcement. Here too will be found the organizations that provide the self-promulgated codes and standards of the private construction industry.

Associations and Professional Groups

ACTION-HOUSING INC., Allegheny Council to Improve Our Neighborhoods, Two Gateway Center, Pittsburgh, Pennsylvania 15222.

AIR POLLUTION CONTROL ASSOCIATION, 4400 Fifth Ave., Pittsburgh, Pennsylvania 15213. Tel.: (412) 621-1100. Works toward international adoption of reasonable engineering performance standards; seeks to establish definitions, methods, processes, procedures, and recommended practical limits of air pollution emissions. Publication:
Journal of the Air Pollution Control Association, monthly

AMERICAN COUNCIL OF INDEPENDENT LABORATORIES, 1725 K St., N.W., Washington, D.C. 20006. Association of independent scientific laboratories. Promotes professional practices. Publication:
ACIL Bulletin, quarterly

AMERICAN INSTITUTE OF PLANNERS, 1776 Massachusetts Ave., N.W., Washington, D.C. 20036. Society of consultants involved in city, county, state, and national planning. Publications:

Journal of the American Institute of Planners, bimonthly
Planners Notebook, bimonthly

AMERICAN INSURANCE ASSOCIATION, 85 John St., New York, N.Y. 10038. Tel.: (212) 433-4400. More than 120 insurance companies through their association promote highway safety, fire prevention, and industrial safety. Publishers of the National Building Code.

AMERICAN NATIONAL STANDARDS INSTITUTE, 1430 Broadway, New York, N.Y. 10018. Tel.: (212) 868-1220. Coordinates and administers the voluntary standardization system in the U.S. More than 550 construction standards are listed in their 1975 catalog.

AMERICAN SOCIETY OF PLANNING OFFICIALS, 1313 E. 60th St., Chicago, Illinois 60637. Provides professional services and information in the fields of zoning, urban renewal, and urban planning. Publications:

Zoning Digest, monthly
Planning Magazine, monthly

BUILDING OFFICIALS AND CODE ADMINISTRATORS INTERNATIONAL, 1313 E. 60th St., Chicago, Illinois 60637. Tel.: (312) 324-3400. Has established a number of building, housing, and mechanical codes and has served local agencies responsible for formulating and enforcing codes. Publication:

The Building Official and Code Administrator, monthly

BUILDING RESEARCH ADVISORY BOARD, 2101 Constitution Ave., N.W., Washington, D.C. 20418. Tel.: (202) 961-1348. Part of the National Academy of Sciences/Engineering, BRAB offers advisory services on housing, building materials, planning, etc., to industry and government. (See Building Research Institute in Section II). Publications:

BRAB-BRI Journal—Building Research, quarterly
BRAB Building Research Institute Newsletter

INTERNATIONAL ASSOCIATION OF ELECTRICAL INSPECTORS, 802 Busse Highway, Park Ridge, Illinois 60068. Tel.: (313) 337-1658. Group of electrical inspectors from government, industry, and insurance firms works with architects, contractors, and others to promote the National Electrical Code. Publication:

IAEI News, bimonthly

INTERNATIONAL ASSOCIATION OF PLUMBING AND MECHANICAL OFFICIALS, 5032 Alhambra Ave., Los Angeles, California 90032. Promotes a uniform plumbing code, which has been adopted by 11 states as a state code and is used by more than 7,000 jurisdictions in 25 states. Publication:

The Official, bimonthly

INTERNATIONAL CITY MANAGEMENT ASSOCIATION, 1140 Connecticut Ave., N.W., Washington, D.C. 20036. Tel.: (202) 293-2200. Develops and disseminates new approaches to management through training programs, information services, and publications. Publication:

Public Management, monthly

INTERNATIONAL CONFERENCE OF BUILDING OFFICIALS, 5360 S. Workman Mill Rd., Whittier, California 90601. Tel.: (213) 699-0541. Publishes and promotes the Uniform Building Code; conducts research in safety. Publication:

Building Standards, bimonthly

NATIONAL ASSOCIATION OF COUNTY PLANNING DIRECTORS, 1001 Connecticut Ave., N.W., Washington, D.C. 20036. Tel.: (202) 628-4701. Publication:

Annual Membership Directory

NATIONAL ASSOCIATION OF HOUSING AND REDEVELOPMENT OFFICIALS, 2600 Virginia Ave., N.W., Washington, D.C. Tel.: (202) 333-2020. Group concerned with public housing, urban renewal, and housing rehabilitation through code enforcement and community action. Publications:

Journal of Housing, 11 times a year
Newsletter, weekly

NATIONAL ASSOCIATION OF HOUSING COOPERATIVES, P.O. Box 5210, Detroit, Michigan 48235. Group of housing cooperatives representing 100,000 families. Publication:
Cooperative Housing Journal

NATIONAL BOARD OF BOILER AND PRESSURE VESSEL INSPECTORS, 1155 N. High St., Columbus, Ohio 43201. Tel.: (614) 294-4957.

NATIONAL CERTIFIED PIPE WELDING BUREAU, 5530 Wisconsin Ave., Suite 750, Washington, D.C. 20015. Tel.: (301) 654-7960. Made up of piping contractors who develop pipe welding standards and procedures.

NATIONAL FIRE PROTECTION ASSOCIATION INTERNATIONAL, 470 Atlantic Ave., Boston, Massachusetts 02210. Members consist of individuals, corporations, firms, municipal departments, and others. Promotes methods of fire protection. Publications:
Fire Journal, bimonthly
Fire News

NATIONAL SAFETY COUNCIL, 425 N. Michigan Ave., Chicago, Illinois 60611. Gathers information and statistics on accidents, how and why they happen, and how they can be prevented. Analyzes and distributes safety information and employs safety engineers who research safety problems and publish solutions.

SOUTHEAST MICHIGAN COUNCIL OF GOVERNMENTS, Book Bldg., Detroit, Michigan 48226. Tel.: (313) 961-4266. Association of counties, cities, villages, townships, and school districts. Works to resolve common problems.

SOUTHERN BUILDING CODE CONGRESS, 3617 8th Ave. S., Birmingham, Alabama 35222. Codifies minimum construction standards. Publication:
Southern Building Magazine

TECHNICO-OP INC., 1010 Washington Blvd., Stamford, Connecticut 06901. Maintains a file of technical bulletins relating facilities and equipment for moderate and lower income housing cooperatives, covering site selection, townhouse design standards, recreation and community building standards, apartment security, design for lower cost insurance, etc.

URBAN LAND INSTITUTE, 1200 18th St., N.W., Washington, D.C. 20036. Tel.: (202) 338-6800. Research and educational organization that works to improve the quality and standards of land use and development. Publications:

Urban Land, monthly
Land Use Digest, monthly
Environmental Comment, monthly

Periodicals

CHICAGOLAND DEVELOPMENT, published monthly by the Chicago Association of Commerce and Industry, 30 W. Monroe St., Chicago, Illinois 60603. Provides data for analysis of land use, housing, industrial and commercial real estate, and construction in the Chicago area; statistical information on building permits, industrial construction (by industry), construction (by county), and mortgage financing. $15 per year for association members; $25 for others.

HUD CHALLENGE, published monthly by the Department of Housing and Urban Development. Provides a medium for discussion official HUD policies, projects, programs, and new directions. Order No. HH 1.36, Government Printing Office, Washington, D.C. 20406. $6.50 annually.

HUD NEWSLETTER, weekly report by the Department of Housing and Urban Development on housing, urban affairs, the mortgage market, credit, and related matters. Order No. HH 1.15/4, Government Printing Office, Washington, D.C. 20406.

JOURNAL OF HOUSING, published 11 months per year by National Association of Housing and Redevelopment Officials, 2600 Virginia Ave., N.W., Washington, D.C. 20037. For staff members and commissioners of the public agencies concerned with slum clearance, low-rent housing, neighborhood conservation, housing code enforcement, and city rebuilding. $8 per year.

SOUTHERN BUILDING, published monthly by Southern Building Code Publishing Co., 1116 Brown-Marx Bldg., Birmingham, Alabama 35203. Information on how new products, methods of construction, and assemblies meet the requirements of the Southern Standard Building Code. No cost to qualified readers; $4 for others.

Directories

ACIL DIRECTORY, published every two years by the American Council of Independent Laboratories, 1725 K St., N.W., Washington, D.C. 20006. Contains alphabetical and geographical lists of members in addition to one-page summaries detailing principals, services, and other facts for each member laboratory.

AMERICAN SOCIETY FOR TESTING AND MATERIALS DIRECTORY OF TESTING LABORATORIES. Lists locations of laboratories equipped to test materials and commodities. American Society for Testing and Materials. For ASTM members, $3; others, $3.75.

BUILDING CONSTRUCTION INFORMATION SOURCES. How to get information on nearly 1,000 building topics. Gale Research Co., Book Tower, Detroit, Michigan 40226. $14.50.

BUILDING OFFICIALS AND CODE ADMINISTRATORS INTERNATIONAL MEMBERSHIP DIRECTORY. Includes name, title, and complete address of each member and representatives, articles of association, and bylaws of the association, and a listing of committees and committee members. Building Officials and Code Administrators International, 1313 E. 60th St., Chicago, Illinois 60637. Cost for members, $4.20; for others, $6.

DEPARTMENT OF DEFENSE INDEX OF SPECIFICATIONS AND STANDARDS. Consolidated edition of indexes of military specifications used by the Army, Navy, and Air Force. Government Printing Office, Washington, D.C. 20402. Order No. D 7.14; $36.

DICTIONARY CATALOG OF THE UNITED STATES DEPARTMENT OF HOUSING AND URBAN DEVELOPMENT LIBRARY AND INFORMATION DIVISION. Catalog of urban and regional data, much of which has been compiled through HUD-sponsored studies. Library contains over 12,000 comprehensive planning reports and Model Cities reports, plus consolidated holdings of the libraries of the Federal Housing Administration, Public Housing Administration, and the Housing and Home Finance Agency. 295,000 cards. G.K. Hall and Co., 70 Lincoln St., Boston, Massachusetts 02111. 1973, 19 volumes, $1,425.

DIRECTORY OF GOVERNMENTAL AIR POLLUTION AGENCIES, published annually by the Air Pollution Control Association, 4400 Fifth Ave., Pittsburgh, Pennsylvania 15213. Listing of governmental agencies—federal, state, and local—dealing with air pollution control. $2.50.

ENVIRONMENT REGULATION HANDBOOK. Summaries, flowcharts, and transcripts of laws and regulations, specimens of permit forms, indexes by federal code and by subject, and retrieval services. Environment Information Center; $95 including updating for one year.

HUD HOUSING AND PLANNING REFERENCES. Contains a selection of publications and articles on housing and planning received in the U.S. Department of Housing and Urban Development Library during a two-month period. Published bimonthly by the Government Printing Office; order No. HH 1.23/3, $9 per year.

INDEX OF FEDERAL SPECIFICATIONS AND STANDARDS. Alphabetical, numerical, and federal supply classification listings of federal specifications. Government Printing Office, Washington, D.C. 20402. Order No. GS 2.8/2:973; $3.50.

INTERNATIONAL ASSOCIATION OF PLUMBING AND MECHANICAL OFFICIALS DIRECTORY OF RESEARCH RECOMMENDATION. Gives results of tests on new construction products and techniques and lists products that meet the Uniform Plumbing Code. International Association of Plumbing and Mechanical Officials, 5032 Alhambra Ave., Los Angeles, California 90032. $25 annually.

MUNICIPAL YEAR BOOK. Contains data, statistical presentations, and interpretive summaries and analyses on U.S. and Canadian cities, counties, and regional councils. Directory section includes the names and phone numbers of local government officials. International City Management Association, 1140 Connecticut Ave., N.W., Washington, D.C. 20036. $22.50.

NAHRA HOUSING DIRECTORY. Listing of housing agencies and key personnel in housing program administrations. National Association of Housing and Redevelopment Officials, 2600 Virginia Ave., N.W., Washington, D.C. $50.

NAHRA HOUSING CODE AGENCY DIRECTORY. State-by-state listing of local agencies that administer local housing codes and federally assisted code enforcement and demolition programs. National Association of Housing and Redevelopment Officials, 2600 Virginia Ave., N.W., Washington, D.C. for NAHRO members, $2; for others, $4.

NAHRO RENEWAL AGENCY DIRECTORY. State-by-state listing of local agencies administering urban renewal programs; gives program size, agency address, name of executive director, and staff size. National Association of Housing and Redevelopment Officials, 2600 Virginia Ave., N.W., Washington, D.C. For NAHRO members, $3.50; others, $7.

NATIONAL DIRECTORY ON HOUSING FOR OLDER PEOPLE. Directory of apartments, mobile home parks, public housing, and retirement communities. National Council on the Aging, 1828 L St., N.W., Washington, D.C. 20036. $5.

OSHA STANDARDS AND REGULATIONS, Vol. III, Construction Standards. Government Printing Office, Washington, D.C. 20402. Order No. L35.6/3-3; $8 annually including updating.

STANDARDS AND SPECIFICATIONS INFORMATION SOURCES. Annotated bibliography in seven sections: general sources and directories, bibliographies and indexes to periodicals, catalogs and indexes of standards and specification, government sources, associations and societies, and periodicals. Gale Research Co., Book Tower, Detroit, Michigan 40226. $14.50.

URBAN RENEWAL HANDBOOK. Official statement of the Department of Housing and Urban Renewal for federally assisted urban renewal programs. Government Printing Office, Washington, D.C. 20402. Order No. HH 1.6/6: RHA 7200.0-7228.1/2. $27 per year.

VI. OTHER CONSTRUCTION INTERESTS—LABOR, LAW, ARBITRATION, EDUCATION, SURVEYING, MANAGEMENT, FINANCE, REAL ESTATE, ETC.

There are many more facets to the construction industry than those covered by the preceding sections. There are, for example, labor unions, real estate appraisers, and schools of construction and architecture. These varied miscellaneous interests are collected here.

Associations and Professional Groups

AMERICAN CONGRESS ON SURVEYING AND MAPPING, 733 15th St., N.W., Washington, D.C. 20005. Tel.: (202) 347-0029. For land surveyors in private practice and employed by local, state, and federal government. Publication:
ACSM Bulletin, quarterly
ACSM Journal, quarterly

AMERICAN FEDERATION OF TECHNICAL ENGINEERS, 1126 16th St., N.W., Washington, D.C. 20036. Tel.: (202) 223-1811. Affiliated with the AFL-CIO. Publication:
Engineers Outlook, monthly

AMERICAN INSTITUTE OF REAL ESTATE APPRAISERS, 155 E. Superior St., Chicago, Illinois 60611. Promotes ethics; conducts and sponsors courses in appraising. Publications:
The Appraiser, ten issues per year
The Appraisal Journal, quarterly

AMERICAN SOCIETY OF APPRAISERS, Box 17265, Dulles International Airport, Washington, D.C. 20041. Tel.: (703) 471-7888. Publications:
Appraisal and Valuation Manual, biannual
Directory, annual

APPRAISERS ASSOCIATION OF AMERICA, 541 Lexington Ave., New York, N.Y. 10022. Tel.: (212) 753-5039. Professional society of real estate appraisers.

ASSOCIATED SCHOOLS OF CONSTRUCTION, West Virginia State College, Institute, West Virginia 25112. Composed of 33 schools offering degrees with major emphasis on construction.

ASSOCIATION OF COLLEGIATE SCHOOLS OF ARCHITECTURE, 1785 Massachusetts Ave., N.W., Washington, D.C. 20036. Group of more than 160 schools that offer a degree in architecture. Publications:

Journal of Architectural Education, quarterly
ACSA Newsletter
Membership Directory

BRICKLAYERS, MASONS, AND PLASTERERS INTERNATIONAL UNION OF AMERICA, 815 15th St., N.W., Washington, D.C. 20005. Tel.: (202) 783-3788. Publication:

Bricklayer, Mason, and Plasterer, monthly

CONSTRUCTION INDUSTRY JOINT CONFERENCE, 1012 14th St., N.W., Washington, D.C. 20005. Tel.: (202) 783-0038. Made up of representatives of the unions affiiliated with the Building and Construction Trades Department of the AFL-CIO, and members of contractors associations.

CONSTRUCTION SURVEYORS INSTITUTE, 420 Lexington Ave., New York, N.Y. 10017. Tel.: (212) 685-7317.

INTERNATIONAL ASSOCIATION OF BRIDGE, STRUCTURAL, AND ORNAMENTAL IRON WORKERS, 1750 New York Ave., N.W., Washington, D.C. 20006. AFL-CIO. Publication:

The Ironworker, monthly

INTERNATIONAL ASSOCIATION OF HEAT AND FROST INSULATORS AND ASBESTOS WORKERS, 1300 Connecticut Ave., N.W., Washington, D.C. 20036. Tel.: (202) 785-2388. AFL-CIO.

INTERNATIONAL ASSOCIATION OF TILE, MARBLE,

AND SLATE AND STONE POLISHERS HELPERS, 821 15th St., N.E., Washington, D.C. 20005. Tel.: (202) 347-7414. AFL-CIO.

INTERNATIONAL BROTHERHOOD OF ELECTRICAL WORKERS, 1125 15th St., N.W., Washington, D.C. 20005. Tel.: (202) 833-7000. AFL-CIO. Publication:
The Electrical Workers Journal, monthly

INTERNATIONAL BROTHERHOOD OF TEAMSTERS, CHAUFFEURS, WAREHOUSEMEN, AND HELPERS OF AMERICA, 25 Louisiana Ave., N.W., Washington, D.C. 20001. Tel.: (202) 783-0525. Publication:
International Teamster, monthly

INTERNATIONAL UNION OF ELEVATOR CONSTRUCTORS, 12 S. 12th St., Philadelphia, Pennsylvania 19107. Tel.: (215) 922-2226. AFL-CIO. Publication:
The Elevator Constructor, monthly

INTERNATIONAL UNION OF OPERATING ENGINEERS, 1125 17th St., N.W., Washington, D.C. 20006. Tel.: (202) 347-8560. AFL-CIO. Publication:
International Operating Engineer, monthly

INTERNATIONAL WOODWORKERS OF AMERICA, 1622 N. Lombard St., Portland, Oregon 97217. Tel.: (503) 285-5281. Publication:
International Woodworker, semimonthly

LABORERS INTERNATIONAL UNION OF NORTH AMERICA, 905 16th St., N.W., Washington, D.C. 20006. AFL-CIO. Publication:
The Laborer, monthly

OPERATIVE PLASTERERS AND CEMENT MASONS INTERNATIONAL ASSOCIATION OF THE U.S. AND CANADA, 1125 17th St., N.W., Washington, D.C. 20036. Tel.: (202) 393-6569. AFL-CIO. Publication:
The Plasterer and Cement Mason, monthly

SHEET METAL WORKERS INTERNATIONAL ASSOCIATION, 1750 New York Ave., N.W., Washington, D.C. 20036. AFL-CIO. Publication:
Sheet Metal Workers Journal, monthly

SOCIETY OF REAL ESTATE APPRAISERS, 7 S. Dearborn St., Chicago, Illinois 60603. Tel.: (312) 346-7422. Publications:
Appraisal Briefs, weekly
The Real Estate Appraiser, bimonthly

UNITED ASSOCIATION OF JOURNEYMEN AND APPRENTICES OF THE PLUMBING AND PIPE FITTING INDUSTRY OF THE U.S. AND CANADA, 901 Massachusetts Ave., N.W., Washington, D.C. 20001. Tel.: (202) 628-5823. AFL-CIO. Publication:
U.A. Journal, monthly

UNITED BROTHERHOOD OF CARPENTERS AND JOINERS OF AMERICA, 101 Constitution Ave., N.W., Washington, D.C. 20001. Tel.: (202) 546-6206. AFL-CIO. Publication:
The Carpenter, monthly

UNITED CEMENT, LIME, AND GYPSUM WORKERS INTERNATIONAL UNION, 7830 W. Lawrence Ave., Chicago, Illinois 60656. Tel.: (312) 774-2217. AFL-CIO. Publication:
Voice of the Cement, Lime, Gypsum, and Allied Workers, monthly

UNITED SLATE, TILE, AND COMPOSITION ROOFERS, DAMP AND WATERPROOF WORKERS ASSOCIATION, 1125 17th St., N.W., Washington, D.C. 20036. Tel.: (202) 638-3228. AFL-CIO. Publication:
Journeyman Roofer and Waterproofer, monthly

WOOD, WIRE, AND METAL LATHERS INTERNATIONAL UNION, 815 16th St., N.W., Washington, D.C. 20006. AFL-CIO. Publication:
The Lather, monthly

Periodicals

AMERICAN INDUSTRIAL PROPERTIES REPORT, published every other month by Indprop Publishing Co., 74 Shrewsbury Ave., Red Bank, N.J. 07701. Covers site selection. Includes geographical classified section of plants and sites available. No cost for qualified readers; $15 per year for others.

ARIZONA PROGRESS, published monthly by the Valley National Bank, Phoenix, Arizona. Report on the Arizona economy, including data on building reports and new housing starts. No charge.

BUILDING SERVICE CONTRACTOR, published monthly by MacNair-Dorland Co., 101 W. 31st St., New York, N.Y. 10001. Covers the contract cleaning field. No cost to qualified readers.

CARPENTER, published monthly by United Brotherhood of Carpenters and Joiners of America, 101 Constitution Ave., Washington, D.C. 20001. $3 per year.

CHICAGOLAND'S REAL ESTATE ADVERTISER, published every Friday by the Law Bulletin Publishing Co., 415 N. State St., Chicago, Illinois 60610. For the real estate and allied fields in the six-county metropolitan area of Chicago. Covers construction, commercial transactions, real estate transfers, mortgage and finance, property management, commercial and residential brokerage, and building developments. No cost to qualified readers; $17.50 per year for others.

CONSTRUCTION LABOR REPORT, published weekly by the Bureau of National Affairs Inc., 1231 25th St., N.W., Washington, D.C. 20037. News of construction labor developments, including wage and salary controls, collective bargaining, union policies, and legislation.

COMMERCIAL RECORD, published weekly by Record Publishing Co., 241 Orange St., P.O. Box 1890, New Haven, Connecticut 06507. Covers real estate, financing, and construction fields in Connecticut. Reports credit information such as bankruptcies, liens, attachments, and foreclosures. $56 per year.

NATIONAL REAL ESTATE INVESTOR, published monthly by Communication Channels Inc., 461 Eighth Ave., New York, N.Y. 10001. News and analyses of the real estate, construction, and financing industries. No cost to qualified readers; $24 per year for others.

NEW ENGLAND REAL ESTATE JOURNAL, published weekly at 36 Washington St., Wellesley Hills, Massachusetts 02181. News of commercial and industrial sites, new construction, financing, lease management agreements, and other developments. $15 per year.

PROPERTIES MAGAZINE, published monthly by Properties Magazine, 4900 Euclid Ave., Cleveland, Ohio 44103. $6 per year.

REAL ESTATE FORUM, published monthly by Real Estate Forum Inc., 39 E. 42nd St., New York, N.Y. Covers new construction, building modernization, law, sales, leases and mortgages, personnel changes, shopping center development, and industrial real estate. No cost to qualified readers.

REAL ESTATE NEWS, published weekly by Real Estate News Inc., 600 W. Van Buren St., Chicago, Illinois 60607. For brokers, salesmen, property managers, appraisers, suppliers, insurers, architects, builders, and financers. $8.50 per year.

REAL ESTATE NEWS (GREATER NEW YORK), published monthly by the Greater New York Taxpayers Association, 770 Broadway, New York, N.Y. 10003.

REALTY, official publication of the National Realty Club, the Bronx Realty Advisory Board, the Bushwick Realty Board of Brooklyn. Published fortnightly by Benenson Publications Inc., 156 E. 42nd St., New York, N.Y. 10022. No cost to qualified readers; $5 per year for others.

REALTY AND BUILDING, published weekly by Realty and Building Inc., 12 E. Grand Ave., Chicago, Illinois 60611. For those in the real estate, construction, architectural, and mortgage-financing fields, and for owners and operators of income-producing residential, commercial, industrial, and office buildings in the metropolitan Chicago area. No cost to qualified readers; $10 per year for others.

SURVEYING AND MAPPING, published quarterly by American Congress on Surveying and Mapping, Suite 430, Woodward Bldg., Washington, D.C. 20005. Covers advances in surveying and mapping; reports on local, regional, national, and international developments in these fields. $12 per year.

TENNESSEE REALTOR, published bimonthly by Tennessee Association of Realtors, 11th Floor, Third National Bank Bldg., Nashville, Tennessee 37219. Articles on association activities, developments affecting the industry in Tennessee, sales features, educational information, and general real estate news. $12 per year.

Directories

AMERICAN INSTITUTE OF REAL ESTATE APPRAISERS DIRECTORY OF MEMBERS. Lists name, address, and phone number for each member. American Institute of Real Estate Appraisers, 155 E. Superior St., Chicago, Illinois 60611. No Charge.

AMERICAN LAND TITLE ASSOCIATION DIRECTORY. Lists members, their company names, addresses, and financial ratings. American Land Title Association, 1828 L St., N.W., Washington, D.C. 20036. $2.

BUILDING CONSTRUCTION EMPLOYERS DIRECTORY. Lists employers associations and building unions in the Chicago Area. Published by Building Construction Employers Association of Chicago, 228 N. LaSalle St., Chicago, Illinois 60601.

BUILDING CONSTRUCTION INFORMATION SOURCES. Gale Research Co., Book Tower, Detroit, Michigan 40226. $14.

NATIONAL REAL ESTATE DIRECTORY, published annually by Communication Channels Inc., 461 Eighth Ave., New York, N.Y. 10001.

NATIONAL ROSTER OF REALTORS DIRECTORY, published annually by Stamats Publishing Co., 427 6th Ave., S.E., Cedar Rapids, Iowa 52406.

REAL ESTATE INFORMATION SOURCES. Subjects include appraisal, brokerage, building, modernization, land development, finance, insurance, urban renewal, land use, condemnation and eminent domain, law. Appendixes list periodicals, associations, government agencies, libraries, sources of forms, tables, maps, films, and recordings. Gale Research Co., Book Tower, Detroit, Michigan 40226. $14.50.